SOLUTIONS MANUAL TO ACCOMPANY

THEORY AND DESIGN FOR MECHANICAL MEASUREMENTS

SECOND EDITION

RICHARD S. FIGLIOLA
Clemson University

DONALD E. BEASLEY
Clemson University

JOHN WILEY & SONS, INC.
New York • Chichester • Brisbane • Toronto • Singapore

Copyright © 1995 by John Wiley & Sons, Inc.

This material may be reproduced for testing or instructional purposes by people using the text.

ISBN 0-471-10422-1

Printed in the United States of America

10 9 8 7 6 5 4 3 2 1

Printed and bound by Malloy Lithographing, Inc.

PROBLEM 1.1

FIND: Explain the term *standard*.

SOLUTION

The term *standard* refers to an object or instrument, a method or a procedure that provides a value of an acceptable accuracy for comparison. A primary standard defines the value of the unit to which it is associated. Secondary standards, while based on the primary standard, are more readily accessible and amenable for use in a calibration. A test standard defines a specific procedure that is to be followed.

PROBLEM 1.2

FIND When should a calibration be performed?

SOLUTION

A calibration should be performed whenever the accuracy level of a measured value must be ascertained. A periodic calibration of measuring instruments serves as a performance check on those instruments and provides a level of confidence in their indicated values.

PROBLEM 1.3

FIND Suggest methods to estimate the accuracy and precision of a dial thermometer.

SOLUTION

Because precision is related to repeatability, a method that repeatedly exposes the instrument to one or more known temperatures could be developed. An important aspect of such a test is to include some mechanism to allow the instrument to change its indicated value following each reading so that it must readjust itself.

For example, we could place the instrument in an environment of constant temperature and note its indicated value and then move the instrument to another constant temperature environment and note its value there. The two chosen temperatures could be representative of the range of intended use of the instrument. By alternating between the two constant temperature environments, differences in indicated values within each environment would be indicative of the precision error to be expected of the instrument at that temperature. Of course, this assumes that the constant temperatures do indeed remain constant throughout the test and the instrument is used in an identical manner for each measurement.

Accuracy requires comparison against a known value, that is a calibration, in order to assess both precision and bias errors. If in the preceding test the temperatures of the two constant temperature environments were known, the above procedure could serve to establish the bias error, as well as precision error of the instrument. The difference between the average of the readings obtained at some known temperature and the known temperature would provide an estimate of the bias error.

PROBLEM 1.4

FIND When is a dynamic calibration needed?

SOLUTION

A dynamic calibration provides information concerning the time dependent performance of an instrument or measuring system. The time and expense of conducting such a calibration would be justified whenever time dependent signals, including sudden changes in input, are to be measured.

PROBLEM 1.5

FIND How does resolution affect accuracy?

SOLUTION

The resolution of a scale is defined by the least significant increment or division on the output display. Resolution affects a user's ability to resolve the information provided at the output display of an instrument or measuring system. Even under static conditions, competent, independent observers might record a different indicated value from a measurement system with a scatter in the recorded data at a level near the resolution of the measurement system. Such data scatter would appear to behave as if brought on by any other system precision error. As such, the output resolution of a measurement system forms a lower limit as to the precision error to be expected. But resolution would not contribute to bias error.

PROBLEM 1.6

FIND How does hysteresis affect accuracy?

SOLUTION

Hysteresis error is the difference between the values indicated by a measurement system when the value measured is increasing in value as opposed to when it is decreasing in value. This despite the fact that the value sensed is the same for either case. During the course of any test, the signal measured could increase or decrease. We would hope that our measurement system would follow such changes. However, because of the effects of hysteresis, the value indicated by the system could depend on the previous value indicated by the system. The use of randomization methods can break up trends incorrectly implied by hysteresis effects. If randomization methods are not used, the hysteresis effect could behave as a bias error.

PROBLEM 1.8

FIND: Identify measurement stages for each device.

SOLUTION

a) thermostat

> Sensor/transducer: bimetallic thermometer
> Output: displacement of thermometer tip
> Controller: mercury contact switch (open:furnace off; closed:furnace on)

b) speedometer

> Sensor: usually a mechanically coupled cable
> Transducer: typically a dc generator that is turned by the cable producing an electrical signal
> Output: typically a pointer/scale (note: often a galvanometer is used to convert the electrical signal in a mechanical rotation of the pointer)

c) microphone/public address system

> Sensor: microphone diaphragm
> Transducer: microphone (converts diaphragm displacement into an electrical signal)
> Signal conditioning: amplifier
> Output: speaker (note: the speaker is a second transducer in this system converting an electrical signal back to a mechanical displacement)

d) stethoscope

    Sensor: diaphragm
    Signal conditioning: coupled diaphragm-connecting tube filters signal and increases signal level
    Output: acoustic wave at ear piece

Note that there is no transducer in this system. It is purely acoustic.

e) audio speaker

    Sensor: coil
    Transducer: coil-magnet-speaker cone that acts as a miniature electrical dc motor
    Output: speaker cone displacement

f) telephone receiver

  Sending end:
    Sensor: diaphragm
    Transducer: diaphragm-magnet-coil that act as a electromechanical generator
    Output: electrical signal
  Receiving end:
    Sensor: coil
    Transducer: coil-magnet-speaker cone that acts as a miniature electrical dc motor
    Output: speaker cone displacement

PROBLEM 1.9

KNOWN: Data of Table 1.5

FIND: input range, output range

SOLUTION

By inspection

$$0.5 \leq x \leq 100 \text{ cm}$$

$$0.4 \leq y \leq 253.2 \text{ V}$$

The input range (x) is from 0.5 to 100 cm. The output range (y) is from 0.4 to 253.2V. The corresponding spans are given by

$$r_i = 99.5 \text{ cm}$$

$$r_o = 252.8 \text{ V}$$

COMMENT

Note that each answer has units shown. By themselves, numerical answers are meaningless. Always show units for data, for each step of data reduction and in all reported results.

PROBLEM 1.10

KNOWN: Data set of Table 1.5

FIND Discuss advantages of different plot formats for this data

SOLUTION:

Both rectangular and log-log plots are shown.

Rectangular grid:
　　An advantage of this format is that is displays the data clearly as having a non-linear relationship. The data trend, while not immediately quantifiable, is established.
　　A disadvantage with this data set is that the poor resolution at low x values makes quantification at low values difficult.

Log-log grid:
　　An advantage of this format with this particular data set is that the data display a linear relationship of the form: $\log y = m \log x + \log b$. This tells us that the data have the relationship, $y = bx^m$. Because of these facts, resolution is equally good over the whole scale.
　　A disadvantage with this format is that one must remember the data has been conditioned to look linear. We are no longer plotting x versus y. This is particularly important to remember when attempting to find the slope of y against x.

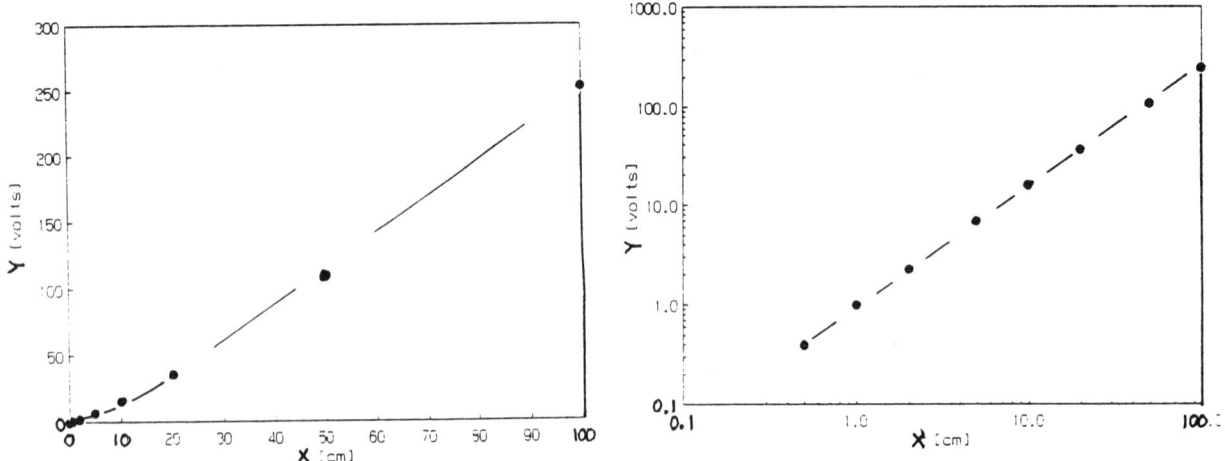

PROBLEM 1.11

KNOWN: Calibration data of Table 1.5

FIND: K at x = 5, 10, 20 cm

SOLUTION:

The data reveal a linear relation on a log-log plot suggesting $y = bx^m$. That is:

$$\log y = \log(bx^m) = \log b + m \log x \quad \text{or}$$
$$Y = B + mX$$

From the plot, B = 0, so that b = 1, and m = 1.2. Thus, we find from the calibration the relationship

$$y = x^{1.2}$$

Because $K = [dy/dx]_x = 1.2 x^{0.2}$, we obtain

x [cm]	K [V/cm]
5	1.66
10	1.90
20	2.18

COMMENT

A common shortcut is to use the approximation that

$$dy/dx = \lim_{x \to 0} \Delta y / \Delta x$$

As given the approximation is valid, but only for very small changes in x. Many students will attempt to use the approximation without consideration of the size of the change in x. Such carelessness will yield erroneous results.

An important aspect of this problem, is to draw attention to the fact that many measurement systems may have a static sensitivity that is dependent on input value. While it is desirable to have a constant K value, the operating principle of many systems will preclude this or incorporate signal conditioning stages to overcome such nonlinearity. In Chapter 3, the concept that system's also have a dynamic sensitivity that is frequency dependent will be introduced.

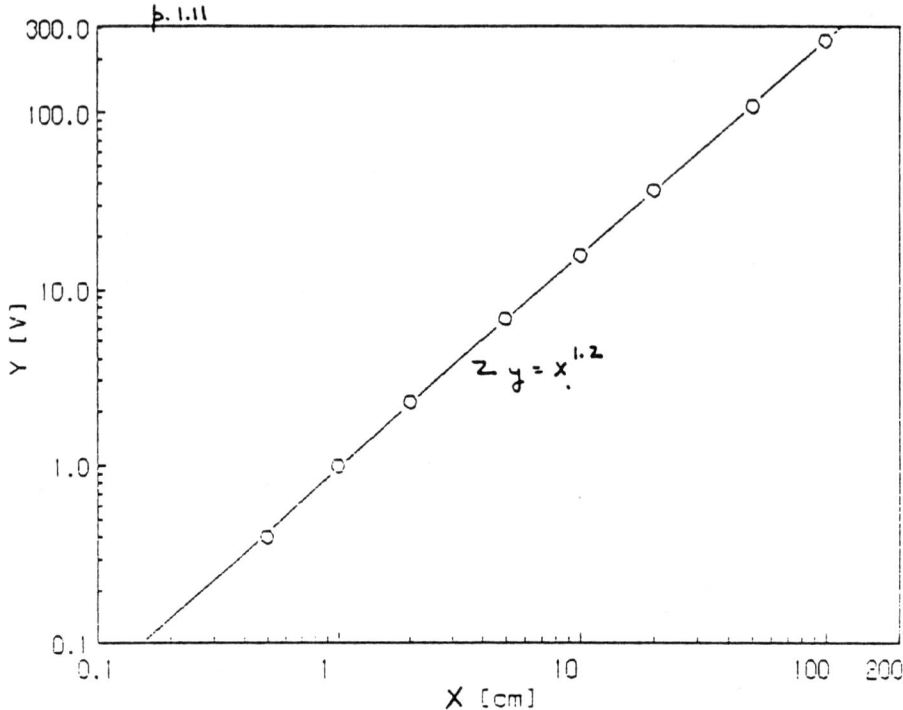

PROBLEM 1.12

KNOWN: Sequence calibration data set of Table 1.6
$r_i = 5$ mV
$r_o = 5$ mV

FIND $\%(e_h)_{max}$

SOLUTION

By inspection of the data, the maximum hysteresis occurs at x = 3.0. For this case,

$$e_h = (e_h)_{max} = y_{up} - y_{down} = 0.2 \text{ mV}$$

or

$$\%(e_h)_{max} = 100 \times (0.2 \text{ mV}/5 \text{ mV}) = 4\%$$

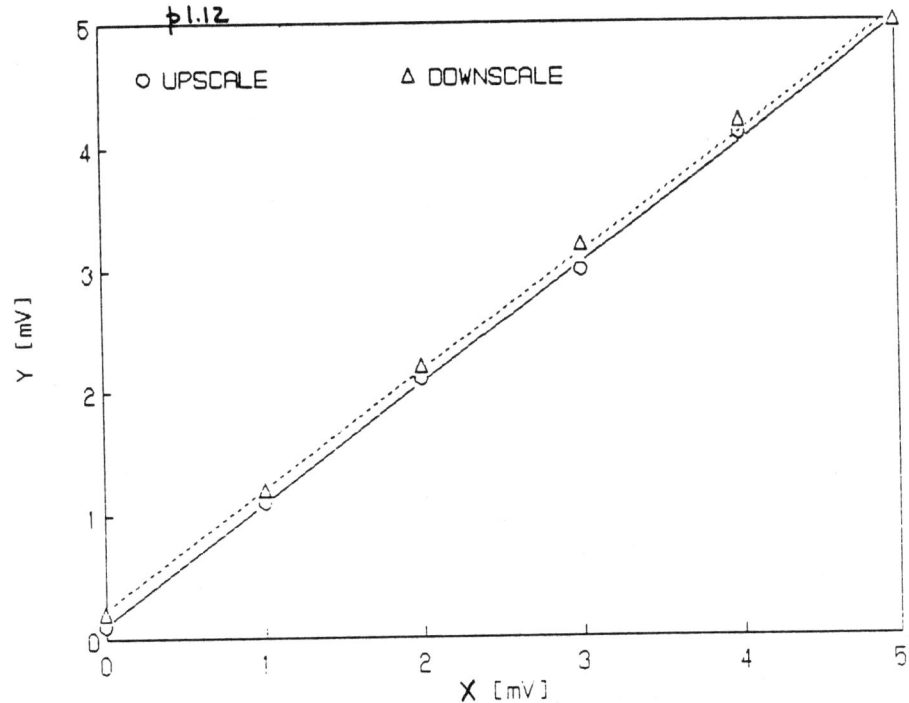

PROBLEM 1.13

KNOWN Comparison of three clock outputs with standard time

FIND Discuss estimated accuracy

SOLUTION

Clock A shows a bias error of 2:23 s. The bias would appear to be increasing at a rate of 1 s/hr. However, clock resolution is 1 s which by itself can lead to precision error (data scatter) of ± 1/2 s; this can create the situation noted here. Another reading would clarify this.

Clock B shows a bias error of 5 s. There does not appear to be any precision error in the output.

Clock C shows a 0 s bias error and a precision error on the order of ± 2 s.

Because of the calibration, we now know the values of bias error for each clock. Correcting for bias error, we can consider Clock B to provide the more accurate time. Over time, the bias error in Clock A could become cumbersome to deal with, that is if the bias is indeed increasing in time. Therefore, it provides the least reliable value of time.

PROBLEM 1.14

SOLUTION

Each curve is plotted below in a suitable format.

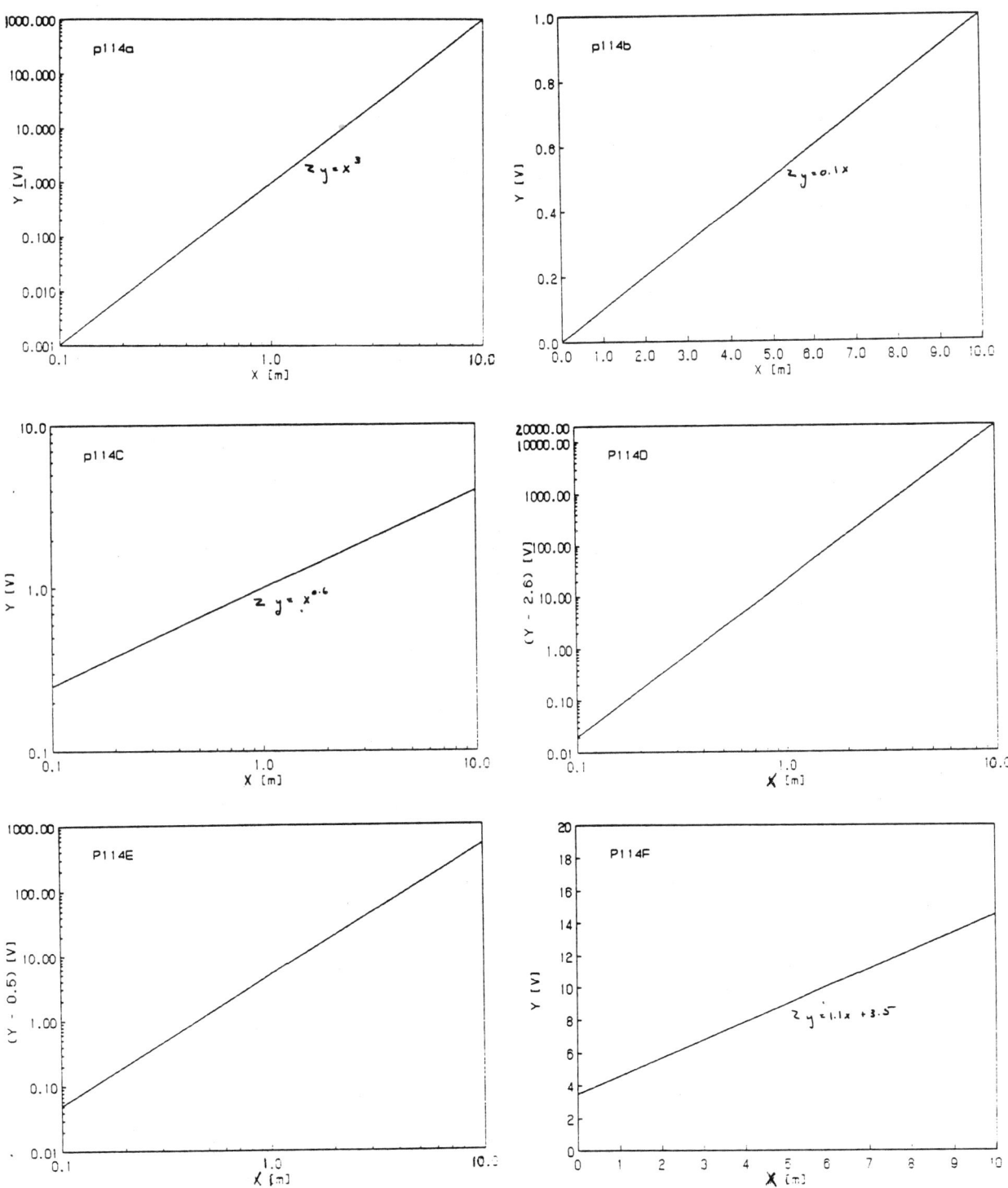

PROBLEM 1.15

KNOWN: $y = 10e^{-5x}$

FIND: Slope at $x = 0, 2$ and 20

SOLUTION

The equation has been plotted below. The slope of the equation at any value of x can be found graphically or by the derivative

$$dy/dx = -50e^{-5x}$$

x [V]	dy/dx [V/unit]
0	-50
2	-0.00227
20	0

The sensitivity of y to x decreases with x.

COMMENT

An important aspect of this problem, is to draw attention to the fact that many measurement systems may have a static sensitivity that is dependent on input value. While it is desirable to have a constant K value, the operating principle of many systems will preclude this or incorporate signal conditioning stages to overcome such nonlinearity. In Chapter 3, the concept that system's also have a dynamic sensitivity that is frequency dependent will be introduced.

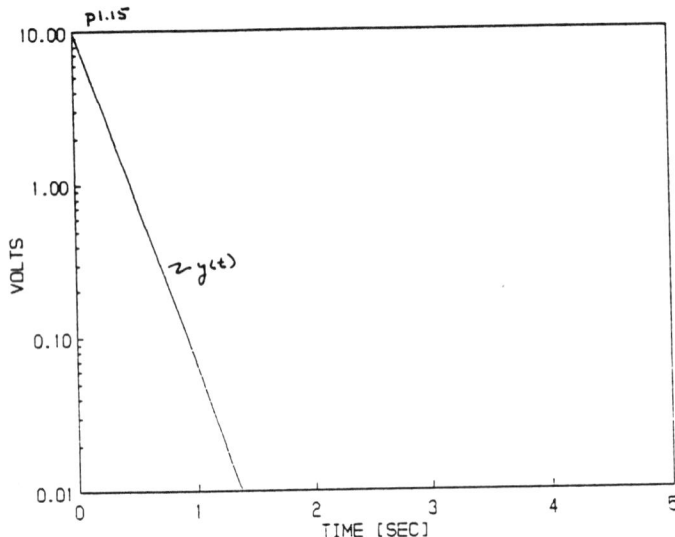

PROBLEM 1.16

SOLUTION

The data are plotted below:

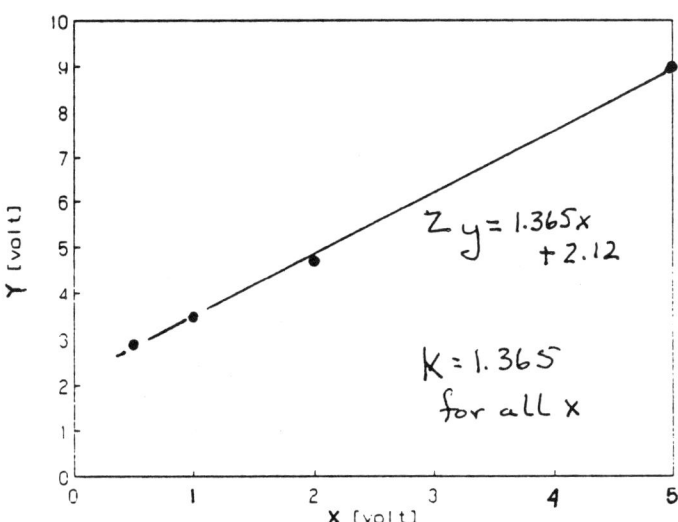

PROBLEM 1.17

KNOWN: Data of form $y = ax^b$.

FIND: a and b; K

SOLUTION

The data are plotted below. If $y = ax^b$, then in log-log format the data will take the linear form

$$\log y = \log a + b \log x$$

A more or less linear curve results with this data. From the plot, the curve fit found is

$$\log y = -0.23 + 2x$$

This implies that

$$y = 0.59x^2$$

so that a = 0.59 and b = 2. The static sensitivity is found by the slope dy/dx at each value of x.

X [m]	$K(X_1) = (dy/dx)\vert_{X_1}$ [cm/m]
0.5	0.54
2.0	2.16
5.0	5.40
10.0	10.80

COMMENT

An important aspect of this problem, is to draw attention to the fact that many measurement systems may have a static sensitivity that is dependent on input value. While it is desirable to have a constant K value, the operating principle of many systems will preclude this or incorporate signal conditioning stages to overcome such nonlinearity. In Chapter 3, the concept that system's also have a dynamic sensitivity that is frequency dependent will be introduced.

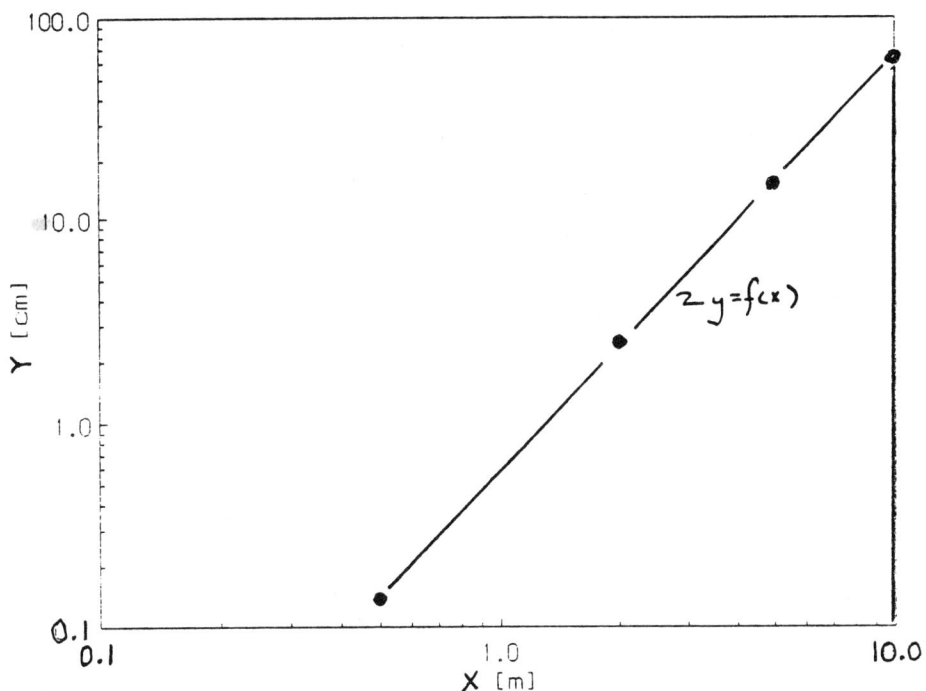

Problem 1.17

PROBLEM 1.18

KNOWN: Calibration data

FIND: Plot. Estimate K.

SOLUTION

The data are plotted below in semi-log format. A linear curve results. This suggests $y = ae^{bx}$. Plotting y vs x in semi-log format is equivalent to plotting

$$\log y = \log a + bx$$

From the plot, a = 5 and b = -1. Hence, the data describe $y = 5e^{-x}$. Now, $K = dy/dx \,|_x$, so that

X [psi]	Y [cm]
0.05	-4.76
0.1	-4.52
0.5	-3.03
1.0	-1.84

The magnitude of the static sensitivity decreases with x. The negative sign indicates that y will decrease as x increases.

COMMENT

An important aspect of this problem is to draw attention to the fact that many measurement systems may have a static sensitivity that is dependent on input value. While it may be desirable to have a constant K value, the operating principle of many systems will preclude this or must incorporate signal conditioning stages to overcome such nonlinearity. In Chapter 3, the concept that systems also have a dynamic sensitivity that is frequency dependent will be introduced.

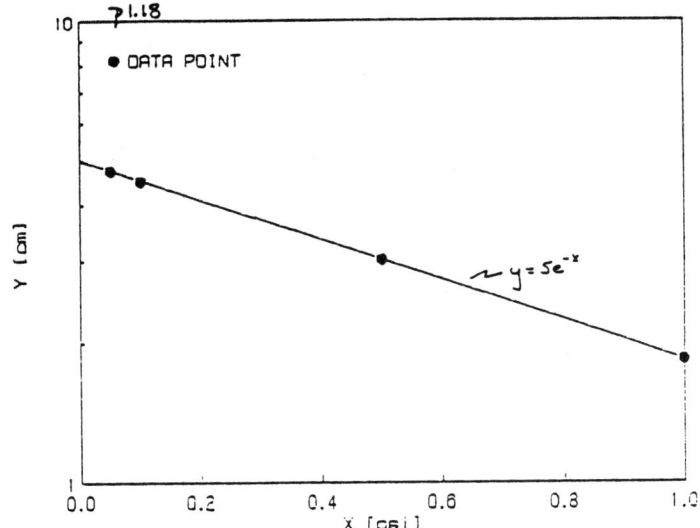

PROBLEM 1.19

KNOWN: A bulb thermometer is used to measure outside temperature.

FIND: Extraneous variables that might influence thermometer output.

SOLUTION

A thermometer's indicated temperature will be influenced by the temperature of solid objects to which it is in contact, and radiation exchange with bodies at different temperatures (including the sky or sun, buildings, people and ground) within its line of sight. Hence, location should be carefully selected and even randomized. We know that a bulb thermometer does not respond quickly to temperature changes, so that a sufficient period of time needs to be allowed for the instrument to adjust to new temperatures. By replication of the measurement, effects due to instrument hysteresis and instrument and procedural repeatability can be randomized.

Because of limited resolution in such an instrument, different competent temperature observers might record different indicated temperatures even if the instrument output were fixed. Either observers should be randomized or, if not, the test replicated. It is interesting to note that such a randomization will bring about a predictable scatter in recorded data of about 1/2 the resolution of the instrument scale.

PROBLEM 1.20

KNOWN: Input voltage, (E_i) and Load (τ_L) can be controlled and varied. Efficiency (η), Winding temperature (T_w), and Current (I) are measured.

FIND: Specify the dependent, independent in the test and suggest any extraneous variables.

SOLUTION

The measured variables are the dependent variables in the test and they depend on the independent variables of input voltage and load. Several influencing extraneous variables include: ambient temperature (T_a) and relative humidity (R); Line voltage fluctuations (e); and each of the individual measuring instruments (m_i). The variation of the independent variables should be performed separately maintaining one independent variable fixed while the other is systematically varied over the test range. A random test procedure for the independent variable will randomize the effects of T_a, R and e. Replication methods using different test instruments would be one way to randomize the effects of the m_i; alternatively, calibration of all measuring instruments would provide a good degree of control over these variables.

$$\eta = \eta(E_i, \tau_L; T_a, R, e, m_i)$$
$$T_w = T_w(E_i, \tau_L; T_a, R, e, m_i)$$
$$I = I(E_i, \tau_L; T_a, R, e, m_i)$$

PROBLEM 1.21

KNOWN: Specifications Table 1.1
Nominal pressure of 500 cm H_2O to be measured.
Ambient temperature drift between 18 to 25 °C

FIND: Magnitude of each elemental error listed.

SOLUTION

Based on the specifications:

r_i = 1000 cm H_2O
r_o = 5 V

Hence, K = 5 V/ 1000 cm H_2O = 5 mV/cm H_2O. This gives a nominal output at 500 cm H_2O input of 2.5 V. This assumes that the input/output relation is linear over range but we are told that it is linear to within some linearity error.

linearity error = e_L = (±0.005) (1000 cm H_2O)
= ± 5 cm H_2O
= ± 0.025 V

hysteresis error = e_h = (±0.0015)(1000 cm H_2O)
= ± 1.5 cm H_2O
= ± 0.0075 V

sensitivity error = e_K = (±0.0025)(500 cm H_2O)
= ± 0.75 cm H_2O
= ± 0.00375 V

thermal sensitivity error = (±0.0002)(7°C)(500 cm H_2O)
= ± 0.7 cm H_2O
= ± 0.0035 V

thermal drift error = (0.0002)(7°C)(1000 cm H_2O)
= 1.4 cm H_2O
= 0.007 V

overall instrument error = $(5^2 + 1.5^2 + 0.75^2 + 0.7^2 + 1.4^2)^{1/2}$ = 5.501 cm H_2O

PROBLEM 1.22

KNOWN: FSO = 1000 N

FIND: e_I

SOLUTION

From the given specifications, the elemental errors are estimated by:

$e_L = 0.001 \times 1000N = 1N$
$e_H = 0.001 \times 1000N = 1N$
$e_R = 0.0015 \times 1000N = 1.5N$
$e_z = 0.002 \times 1000N = 2N$

The overall instrument error is estimated by:

$e_I = (1^2 + 1^2 + 1.5^2 + 2^2)^{1/2} = 2.9N$

COMMENT

This root-sum-square (RSS) method provides a "probable" estimate (i.e. the most likely estimate) of the instrument error possible in any given measurement. "Possible" is the big word here as error values will most likely change between measurements.

PROBLEM 1.23

SOLUTION

Repetition through repeated measurements made under a fixed set of operating conditions provides a measure of the time (or spatial) variation of a measured variable.

Replication through the duplication of tests conducted under similar operating conditions provides a measure of the effect of control of the operating conditions on the measured variable.

Repetition refers to repeating the measurement during a test. Replication refers to repeating the test (to repeat the measurements).

PROBLEM 1.24

SOLUTION

Replication is used to assess the ability to control any aspect of a test or its operating condition. Repeat the test resetting the operating conditions to their original set points.

PROBLEM 1.25

SOLUTION

Randomization is used to break-up the effects of interference from either continuous or discrete extraneous (i.e. uncontrolled) variables.

PROBLEM 1.27

FIND: Test plan to correlate thermostat setting with average room setting

SOLUTION

Although there is no single test plan, one method of solution is as follows.

Assume that average room temperature, T, is a function of actual thermostat setting, spatial distribution of temperature, temporal temperature distribution, and thermostat location. We might imagine that for a controlled (fixed) thermostat location, a direct correlation between setting and T could be achieved. However, factors could influence the temperature measured by the thermostat such as sunlight directly hitting the thermostat or the wall on which it is attached or a location directly exposed to furnace forced convection, a condition aggrevated by air conditioners or heat pumps in which delivered air temperature is a strong function of outside temperature. Assume a proper location is selected and controlled.

Further, the average room temperature must be defined because local room temperature will vary will position within the room and with time. For the test plan, the room should be divided into equal areas with temperature sensing devices placed at the center of each area. The output from each sensor will be averaged over a time period that is long compared to the typical furnace on/off cycle.

Select four temperature sensors: A, B, C, D. Select four thermostat settings: s_1, s_2, s_3, s_4, where $s_1 < s_2 < s_3 < s_4$. Temperatures are to be measured under each setting after the room has adjusted to the new setting. One plan might be:

BLOCK

1	s_1: A, B, C, D
2	s_4: A, B, C, D
3	s_3: A, B, C, D
4	s_2: A, B, C, D

Note that the order of set temperature has been shuffled to attempt to randomize the test plan (hysteresis is a common problem in thermostats). The four blocks will yield the average temperatures, T_1, T_4, T_3, T_2. The data can be presented in a form of T versus s.

PROBLEM 1.28

FIND: Test plan to evaluate fuel efficiency of a production model of automobile

ASSUMPTIONS: Automobile model design is fixed (i.e. neglect options). Require representative estimate of efficiency.

SOLUTION

Although there is no single test plan, one method of solution is as follows. Many variables can affect auto model efficiency: e.g. individual car, driver, terrain, speed, ambient conditions, engine model, fuel, tires, options. Whether these are treated as controlled variables or as extraneous variables depends on the test plan. Suppose we "control" the options, fuel, tires, and engine model, that is fix these for the test duration. Furthermore, we can fix the terrain and the ambient conditions by using a mechanical dynamometer in an enclosed, controlled environment. In fact, such a machine and test conditions have been specified within the U.S.A. by government test standards. By programming the dynamometer to start, accelerate and stop using a preprogrammed routine, we can eliminate the effects of different drivers on different cars. However, this test will fail to randomize the effects of different drivers and terrain as noted in the government statement "... these figures may vary depending on how and where you drive ... ." This leaves the car itself and the test speed as independent variables, x_a and x_b, respectively. We defer considering the effects of the instruments and methods used to compute fuel efficiency until a later chapter, but assume here that this can be done with sufficient accuracy.

With this in mind, we could choose three representative cars and three speeds with the test plan:

BLOCK

1   x_{a1}: x_{b1}, x_{b2}, x_{b3}
2   x_{a2}: x_{b1}, x_{b2}, x_{b3}
3   x_{a3}: x_{b1}, x_{b2}, x_{b3}

Note that since slight differences will exist between cars that can not be controlled, the autos are treated as extraneous variables. This plan randomizes the effects of differences between cars at three different speeds and yields a curve for fuel efficiency versus speed.

As an alternative, we could introduce a driver into the plan. We could develop a test track of fixed (controlled) terrain. And we could have three drivers drive three cars at three different speeds. This introduces the driver as an extraneous variable, noted as A_1, A_2 and A_3 for each driver. Assuming that the tests are run under similar ambient conditions, one test plan may be

	x_{a1}	x_{a2}	x_{a3}
A_1	x_{b1}	x_{b2}	x_{b3}
A_2	x_{b2}	x_{b3}	x_{b1}
A_3	x_{b3}	x_{b1}	x_{b2}

PROBLEM 1.29

KNOWN: Four lathes, 12 machinists are available to produce batches of machine shafts.

FIND: Test plan to estimate the tolerances held within a batch

SOLUTION

If we assume that batch precision, P, is only a function of lathe and machinist, then

$$P = f(\text{lathe, machinist})$$

We can set up a test plan using all four lathes, L_1, L_2, L_3, L_4, and all 12 machinists, A, B, ..., L. The machinists are randomly assigned.

BLOCK

1	L_1: A, B, C
2	L_2: D, E, F
3	L_3: G, H, I
4	L_4: J, K, L

Data from each lathe should be indicative of the precision associated with each lathe and the total ensemble of data indicative of batch precision. However, this test plan neglects the effects of shift and day of the week.

One method which treats machinist and lathe as extraneous variables and reduces test size selects 4 machinists at random. Suppose more than one shaft size is produced at the plant. We could select 4 shaft diameters, D_1, D_2, D_3, D_4 and set up a Latin square:

	L_1	L_2	L_3	L_4
B	D_1	D_2	D_3	D_4
E	D_2	D_3	D_4	D_1
G	D_3	D_4	D_1	D_2
L	D_4	D_1	D_2	D_3

Note that neither plan includes shift or day of the week effects.

PROBLEM 1.30

SOLUTION

Linearity error

A random static calibration over a specified range will provide the input-output relationship between y and x (i.e. $y = f(x)$). A first-order curve fit to this data, for example using a least squares regression analysis, will provide the fit $y_L(x)$. The linearity error is simply the difference between the measured value of y at any value of x and the value of y_L predicted by the fit at that x.

A manufacturer may wish to keep the linearity error below some target value and, hence, may limit the recommended operating range for the system for this purpose. In your experience, you may notice that some systems can be operated outside of their specification range but be aware their elemental errors may exceed the manufacturer's stated values.

Hysteresis error

A sequential static calibration over a specified range will provide the input-output behavior between y and x during upscale-only and downscale-only operations. This will tend to maximize any hysteresis in the system. The hysteresis error is the difference between the upscale value and the downscale value of y at any given x.

PROBLEM 1.31

KNOWN: 4 brands of tires
8 cars of the same make

FIND: Test plan to evaluate performance

SOLUTION

Tire performance can mean different things but for passenger tires usually refers to braking and lateral load adhesion during wet and dry operations. For a given series of performance tests, performance will depend on tire and car (a tire will perform differently on different makes of cars). For the same make, subtle differences in production models can affect test results so we treat the car as an individual and extraneous variable.

We could select 4 cars at random (1,2,3,4) to test four tire brands (A,B,C,D)

        1: A, B, C, D
        2: A, B, C, D
        3: A, B, C, D
        4: A, B, C, D

This provides a data pool for evaluating tire performance for a make of car. Note we ignore the effect of the test driver but this method will incorporate driver variation in by testing four cars. Other strategies could be created.

PROBLEM 1.32

KNOWN: Water at 20°C
Q = f(C,A,dp,ρ)
C = 0.75; D = 1 m
2 < Q < 10 cmm

FIND: Expected calibration curve

SOLUTION

Part of a test plan is to specify the range of the independent variable and to anticipate the range resulting in the dependent variable. In this case, the pressure drop will be measured so that it is the dependent variable during a static calibration. To anticipate the output range of the calibration then:

Rearranging the known relation,

$$dp = (Q/CA)^2 \rho/2$$

For ρ = 998 kg/m³ (Appendix C), and A = $\pi D^2/4$, we find:

Q (cmm)	dp (N/m²)
2	1.6
5	10
10	40

This is plotted below. It is clear that K will not be a constant as K = f(Q).

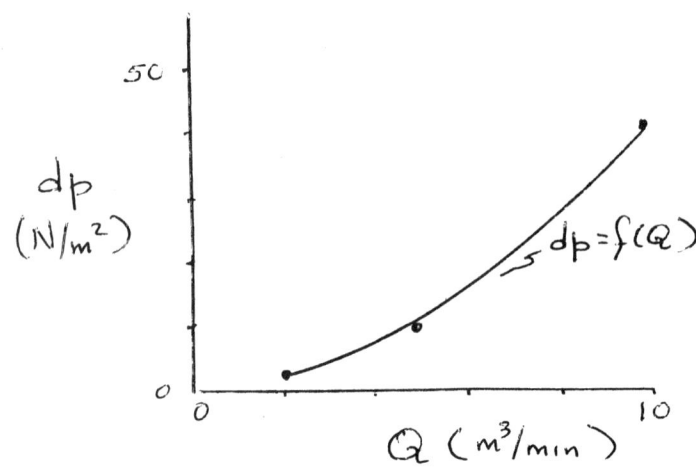

PROBLEM 1.33

SOLUTION

Obviously because $Q \propto dp^{1/2}$ is not linear, the calibration can not be linear. The term can not be applied directly.

However, a signal conditioning stage can be inserted within the signal path to produce a linear output. This is done using logarithmic amplifiers. To illustrate this, plot the calibration curve in Problem 1.32 on a log-log scale. The result will be a linear curve. A linearity measure can be extracted.

PROBLEM 1.34

KNOWN: Pistons are sent out for plating
Four subcontractors

FIND: Test plan for quality control

SOLUTION

Consider four subcontractors as A, B, C, D. One approach is to number the pistons and allocate them to the four subcontractors with subsequent analysis of the plating results. For example, send 24 pistons each to the four subcontractors and analyze the resulting products separately. The variation for each subcontractor can be estimated and can be statistically tested for significant differences.

PROBLEM 1.35

SOLUTION

Controlled variables
    A and B (i.e. control the materials of two alloys)
    T_2 (reference junction temperature)

Independent variable
    T_1 (measured temperature)

Dependent variable
    E (output voltage measured)

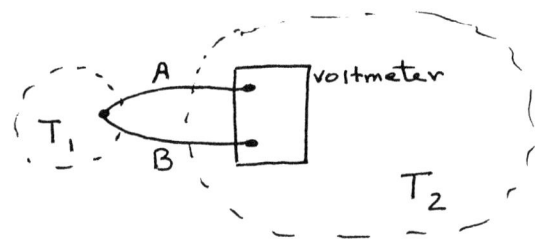

PROBLEM 1.36

SOLUTION

Independent variables
    micrometer setting (i.e. the applied displacement)

Controlled variable
    power supply input

Dependent variable
    output voltage measured

Extraneous variables
    operator set-up, zeroing of system, and reading of micrometer
    ability to set control variables

COMMENT

If you try this you will find that the ability to control the power supply set point can have a significant influence on the results. Not the variations in power during the test, but rather the ability to provide the exact voltage on replication is important. If you use a regular lab variable power supply (rather than a fixed supply), this effect will dominate your data variation on replication.

PROBLEM 1.37

SOLUTION

To test for repeatability, displace the core to various random values over a selected range, such as its expected range, and develop a data base. Data scatter about a curve fit will provide a measure of repeatability for this instrument (methods are discussed in Chapter 4).

Reproducibility involves retesting the system at a different facility. Even though a similar procedure and test plan will be used, the duplication involves different individual instruments and test fixtures. This provides for replication. The combined results allow for interference effects to be randomized.

Bottom Line: The results leading to a reproducibility specification are more representative of what can be expected by the end user (YOU!).

PROBLEM 1.38

SOLUTION

Independent variable
   Applied tensile load

Controlled variable
   Bridge excitation voltage

Dependent variable
   Bridge output voltage (which is related to gauge resistance changes due to
   the applied load)

Extraneous variables
   Specimen and ambient temperature will affect gauge resistance

A replication will involve resetting the control variable and specimen and duplicating the test.

PROBLEM 1.39

SOLUTION

To test repeatability, apply various tensile loads at random over the useful operating range of the system to build a data base. Be sure to operate within the elastic limit of the specimen. Direct comparison and data scatter about a curve fit will provide a measure of repeatability (specific methods to evaluate this are discussed in C4).

Reproducibility involves retesting the system at a different facility. Even though a similar procedure and test plan will be used, the duplication involves different individual instruments and test fixtures. This provides for replication. The combined results allow for interference effects to be randomized.

Bottom Line: The results leading to a reproducibility specification are more representative of what can be expected by the end user (YOU!).

PROBLEM 2.1

FIND: Define signal and provide examples of static and dynamic input signals to measurement systems.

SOLUTION: A signal is information in motion from one place to another, such as between stages of a measurement system. Signals may be in a variety of forms, including electrical and mechanical.

Examples of static signals are:

1. weight, such as weighing merchandise, etc.
2. body temperature, over the time period of interest
3. length or height, such as the length of a board or a person's height

Examples of dynamic signals:

1. input to an automobile speed control
2. input to a stereo amplifier from a component
3. input signal to a printer from a computer

PROBLEM 2.2

FIND: List the important characteristics of signals and define each.

SOLUTION:

1. Magnitude - generally refers to the maximum value of a signal
2. Range - difference between maximum and minimum values of a signal
3. Amplitude - indicative of signal fluctuations relative to the mean
4. Frequency - describes the time variation of a signal
5. Dynamic - signal is time varying
6. Static - signal does not change over the time period of interest
7. Deterministic - signal can be described by an equation
(other than a Fourier series or integral approximation)
8. Non-deterministic - describes a signal which has no discernible pattern of repetition and cannot be described by a simple equation.

PROBLEM 2.3

KNOWN:

$$y(t) = 30 + 2\cos 6\pi t$$

FIND: $\bar{y}$ and y_{rms} for the time periods t_1 to t_2 listed below

    a) 0 to 0.1 s
    b) 0.4 to 0.5 s
    c) 0 to 1/3 s
    d) 0 to 20 s

SOLUTION:

For the function y(t)

$$\bar{y} = \frac{1}{t_2 - t_1} \int_{t_1}^{t_2} y(t)\,dt$$

and

$$y_{rms} = \sqrt{\frac{1}{t_2 - t_1} \int_{t_1}^{t_2} [y(t)]^2\,dt}$$

Thus in general,

$$\bar{y} = \frac{1}{t_2 - t_1} \int_{t_1}^{t_2} (30 + 2\cos 6\pi t)\,dt$$

$$= \frac{1}{t_2 - t_1} \left[30t + \frac{2}{6\pi} \sin 6\pi t \right]_{t_1}^{t_2}$$

$$= \frac{1}{t_2 - t_1} \left[30(t_2 - t_1) + \frac{2}{6\pi} (\sin 6\pi t_2 - \sin 6\pi t_1) \right]$$

and

$$y_{rms} = \left\{ \frac{1}{t_2 - t_1} \int_{t_1}^{t_2} (30 + 2\cos 6\pi t)^2 \,dt \right\}^{1/2}$$

$$= \left\{ \frac{1}{t_2 - t_1} \left[900t + \frac{120}{6\pi} \sin 6\pi t + 4\left(\frac{1}{12\pi} \sin 6\pi t \cos 6\pi t + \frac{1}{2}t \right) \right]_{t_1}^{t_2} \right\}^{1/2}$$

The resulting values are

a) $\bar{y} = 31.01$ $y_{rms} = 31.02$

b) $\bar{y} = 28.99$ $y_{rms} = 29.00$

c) $\bar{y} = 30$ $y_{rms} = 30.03$

d) $\bar{y} = 30$ $y_{rms} = 30.03$

COMMENT: The average and rms values for the time period 0 to 20 seconds represents the long-term average behavior of the signal. The values which result in parts a) and b) are accurate over the specified time periods, and for a measured signal may have specific significance. The period 0 to 1/3 represents one complete cycle of the simple periodic signal and results in average and rms values which accurately represent the long-term behavior of the signal.

PROBLEM 2.4

KNOWN: Discrete, sampled data, corresponding to measurement every 0.4 seconds.

FIND: The mean and rms values of the measured data.

SOLUTION:

The mean value for y_1 is 0 and for y_2 is also 0.
However, the rms value of y_1 is 13.49 and for y_2 is 17.53.

COMMENT: The mean value contains no information concerning the time varying nature of a signal; both these signals have an average value of 0. But the differences in the signals is made apparent when the rms value is examined.

PROBLEM 2.5

KNOWN: The effect of a moving average signal processing technique is to be determined for the signal in Figure 2.22 and y(t) = sin 5t + cos 11t.

FIND: Discuss Figure 2.23 and plot the signal resulting from applying a moving average to y(t).

ASSUMPTIONS: The signal y(t) may be represented by making a discrete representation with Δt equal to 0.05.

SOLUTION:

a) The signal in Figure 2.23 clearly has a reduced level of high frequency content. In essence, this emphasizes longer term variations, while removing short-term fluctuations. It is clear that the peak-to-peak value in the original signal is significantly higher than in the signal which has been averaged.

b) These figures show the effects of a moving average applied to y(t).

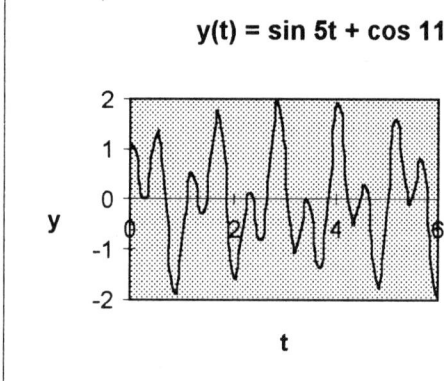

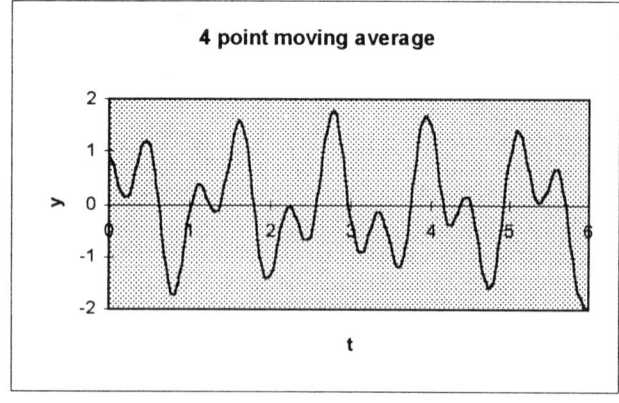

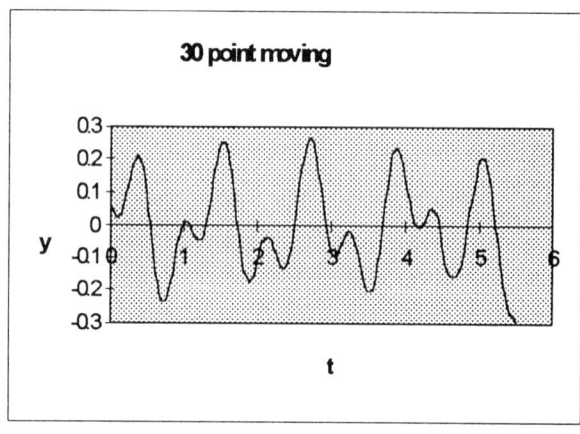

PROBLEM 2.6

KNOWN: A spring-mass system, with

$$m = 0.5 \text{ kg}$$
$$T = 1 \text{ s}$$

FIND: Spring constant, k, and natural frequency ω

SOLUTION:

Since

$$\omega = \sqrt{\frac{k}{m}}$$

(as shown in association with equation 2.7)

and

$$T = \frac{2\pi}{\omega} = \frac{1}{f} = 1 \text{ s}$$
$$\omega = 6.283 \text{ rad/s}$$

The natural frequency is then found as $\omega = 6.283$ rad/s

and
$$\omega = 6.283 = \sqrt{\frac{k}{0.5 \text{ kg}}}$$
$$k = 19.74 \text{ N/m (kg/sec}^2\text{)}$$

PROBLEM 2.7

KNOWN: A spring-mass system having

$$m = 1 \text{ kg}$$
$$k = 5000 \text{ N/cm}$$

FIND: The natural frequency in rad/sec and Hz (ω and f)

SOLUTION:

The natural frequency may be determined,

$$\omega = \sqrt{\frac{k}{m}} = \sqrt{\frac{5000 \frac{N}{cm} \cdot 100 \frac{cm}{m}}{1 \text{ kg}}} = 707.1 \text{ rad/s}$$

and

$$f = \frac{\omega}{2\pi} = 112.5 \text{ Hz}$$

PROBLEM 2.8

KNOWN: Functions:

a) $\sin 10t$
b) $\cos \frac{2\pi t}{3}$
c) $\sin 3n\pi t$ for $n = 1$ to ∞

FIND: The period, frequency in Hz, and circular frequency in rad/s.

SOLUTION:

a) $\omega = 10$ rad/s $f = 1.592$ Hz $T = 0.628$ s

b) $\omega = 2\pi/3$ rad/s $f = 1/3$ Hz $T = 3$ s

c) $\omega = 3n\pi$ rad/s $f = 3n/2$ Hz $T = 2/(3n)$ s

PROBLEM 2.9

KNOWN: $y(t) = 10 \sin 2t + 3 \cos 2t$

FIND: Equivalent expression containing a cosine term only

SOLUTION: From Equation 2.10 and 2.11

$$y = C\cos(\omega t - \phi) \quad \phi = \tan^{-1}\frac{B}{A}$$

and with

$$A\cos\omega t + B\sin\omega t = \sqrt{A^2 + B^2}\cos(\omega t - \phi)$$

we find

$$C = \sqrt{A^2 + B^2} = \sqrt{10^2 + 3^2} = 10.44$$

$$\phi = \tan^{-1}\frac{10}{3} = 1.279 \text{ rad}$$

and

$$y = 10.44\cos(2t - 1.279)$$

PROBLEM 2.10

KNOWN: $y(t) = 6 \sin 4\pi t + 28 \cos 4\pi t$

FIND:

a) Equivalent expression containing a cosine term only
b) Equivalent expression containing a sine term only

SOLUTION: From Equation 2.10 and 2.11

$$y = C\cos(\omega t - \phi) \quad \phi = \tan^{-1}\frac{B}{A}$$

$$y = C\sin(\omega t + \phi^*) \quad \phi^* = \tan^{-1}\frac{A}{B}$$

and with

$$A\cos\omega t + B\sin\omega t = \sqrt{A^2 + B^2}\cos(\omega t - \phi)$$
$$A\cos\omega t + B\sin\omega t = \sqrt{A^2 + B^2}\sin(\omega t + \phi^*)$$

we find

$$C = \sqrt{A^2 + B^2} = \sqrt{28^2 + 6^2} = 28.64$$

$$\phi = \tan^{-1}\frac{6}{28} = 0.21 \text{ rad}$$

$$\phi^* = \tan^{-1}\frac{28}{6} = 1.36 \text{ rad}$$

and

$$y = 28.64\cos(4\pi t - 0.21) \quad \text{Answer (a)}$$
$$y = 28.64\sin(4\pi t + 1.36) \quad \text{Answer (b)}$$

PROBLEM 2.11

KNOWN: $y(t) = \sum_{n=1}^{\infty} \frac{2\pi n}{6} \sin n\pi t + \frac{4\pi n}{6} \cos n\pi t$

FIND:

a) Equivalent expression containing a cosine term only

SOLUTION: From Equation 2.19

$$y = \sum_{n=1}^{\infty} C_n \cos(\omega t - \phi_n) \quad \phi_n = \tan^{-1} \frac{B_n}{A_n}$$

with

$$C_n = \sqrt{A_n^2 + B_n^2}$$

and with

$$C_n = \sqrt{\left(\frac{2\pi n}{6}\right)^2 + \left(\frac{4\pi n}{6}\right)^2} = \sqrt{\frac{5}{9}}\pi n \quad \text{and } \phi_n = \tan^{-1} \frac{(2\pi n/6)}{(4\pi n/6)} = \tan^{-1} 0.5 = 0.46 \text{ rad}$$

we find

$$y(t) = \sum_{n=1}^{\infty} \sqrt{\frac{5}{9}}\pi n \cos(n\pi t - 0.46)$$

PROBLEM 2.12

KNOWN: T is a period of y(x)

FIND: Show that nT for n=2,3,... is a peiod of y(x)

SOLUTION: Since T is a period of y(x)

$$y(x + T) = y(x)$$

Letting $x_1 = x + T$ yields

$$y(x_1) = y(x) = y(x + T)$$

But since T is a period of y(x)

$$y(x + 2T) = y(x_1 + T) = y(x)$$

By analogy then

$$y(x + nT) = y(x)$$

PROBLEM 2.13

KNOWN:

$$y(t) = \sum_{n=1}^{\infty} \frac{3n}{2}\sin nt + \frac{5n}{3}\cos nt$$

FIND: a) fundamental frequency and period
b) cosine series

SOLUTION:

a) The fundamental frequency corresponds to n=1, so $\omega = 1$ rad/s; $T = 2\pi$

b) From equation 2.19

$$y(t) = A_o + \sum_{n=1}^{\infty} C_n \cos\left(\frac{2n\pi t}{T} - \phi_n\right)$$

$$C_n = \sqrt{A_n^2 + B_n^2} \qquad \tan\phi_n = \frac{B_n}{A_n}$$

For this Fourier series

$$C_n = \sqrt{\left(\frac{3n}{2}\right)^2 + \left(\frac{5n}{3}\right)^2} = \sqrt{\frac{181}{36}n^2}$$

$$\phi_n = \tan^{-1}\left(\frac{9}{10}\right) \Rightarrow \phi = 0.7328$$

Thus the series may be written

$$y(t) = \sum_{n=1}^{\infty} \sqrt{\frac{181}{36}n^2}\cos(nt - 0.7328)$$

PROBLEM 2.14

KNOWN:

$$y(t) = 4 + \sum_{n=1}^{\infty} \frac{2n\pi}{10} \cos \frac{n\pi}{4} t + \frac{120 n\pi}{30} \sin \frac{n\pi}{4} t$$

FIND:

a) ω_1 and f_1 b) T_1

c) $y(t) = A_o + \sum_{n=1}^{\infty} C_n \sin\left(\frac{2n\pi t}{T} + \phi^*\right)$

SOLUTION: a) When n=1, $\omega_1 = \pi/4$, $f_1 = 1/8$

b) $T_1 = 8s$

c) From (2.21)

$$C_n = \sqrt{A_n^2 + B_n^2} \quad \text{and} \quad \tan \phi^* = \frac{A_n}{B_n}$$

which yields for this Fourier series

$$C_n = \sqrt{\left(\frac{2n\pi}{10}\right)^2 + \left(\frac{120 n\pi}{30}\right)^2} = 4n\pi$$

$$\phi_n^* = \tan^{-1} \frac{(2n\pi/10)}{(120 n\pi/30)} = \tan^{-1}\left(\frac{1}{20}\right) = 0.05$$

and the Fourier sine series is

$$y(t) = 4 + \sum_{n=1}^{\infty} 4n\pi \sin\left(\frac{n\pi}{4} t + 0.05\right)$$

PROBLEM 2.15

KNOWN:

$$y(t) \begin{array}{ll} = 0 & \text{for } -\pi \leq t \leq 0 \\ = -1 & \text{for } 0 \leq t \leq \pi/2 \\ = 1 & \text{for } \pi/2 \leq t \leq \pi \end{array}$$

FIND: Fourier series for y(t)

SOLUTION: Since the function is neither even nor odd, the Fourier series will contain both sine and cosine terms. The coefficients are found as

$$A_o = \frac{1}{T}\int_{-T/2}^{T/2} y(t)dt = \frac{1}{2\pi}\int_{-\pi}^{\pi} y(t)dt =$$

$$\frac{1}{2\pi}\left[\int_{-\pi}^{0} 0\,dt + \int_{0}^{\pi/2} -1\,dt + \int_{\pi/2}^{\pi} 1\,dt\right] =$$

$$\frac{1}{2\pi}\left[\left(-\pi/2 - 0\right) + \left(\pi - \pi/2\right)\right] = 0 \therefore A_o = 0$$

Note: Since the contribution from $-\pi$ to 0 is identically zero, it will be omitted.

$$A_n = \frac{2}{2\pi}\left[\int_{0}^{\pi/2} -1\cos\frac{2n\pi t}{T} + \int_{\pi/2}^{\pi} 1\cos\frac{2n\pi t}{T}\right]dt$$

$$= \frac{1}{\pi}\left\{\left[\frac{-1}{n}\sin nt\right]_0^{\pi/2} + \left[\frac{1}{n}\sin nt\right]_{\pi/2}^{\pi}\right\}$$

$$= \frac{-1}{\pi n}2\sin\left(\frac{n\pi}{2}\right)$$

$$B_n = \frac{2}{2\pi}\left[\int_{0}^{\pi/2} -1\sin\frac{2n\pi t}{T}dt + \int_{\pi/2}^{\pi} 1\sin\frac{2n\pi t}{T}dt\right]$$

$$= \frac{1}{n\pi}\left\{[\cos nt]_0^{\pi} + [-\cos nt]_{\pi/2}^{\pi}\right\} = \frac{1}{n\pi}[-\cos(0) - \cos(n\pi)]$$

Noting that A_n is zero for n even, and B_n is zero for n odd, the resulting Fourier series is

$$y(t) = \frac{2}{\pi}\left[-\cos t - \frac{1}{2}\sin 2t + \frac{1}{3}\cos 3t - \frac{1}{4}\sin 4t - \frac{1}{5}\cos 5t - \frac{1}{6}\sin 6t + \frac{1}{7}\cos 7t - \ldots\right]$$

Partial sums of this Fourier series and the function y(t) are plotted below.

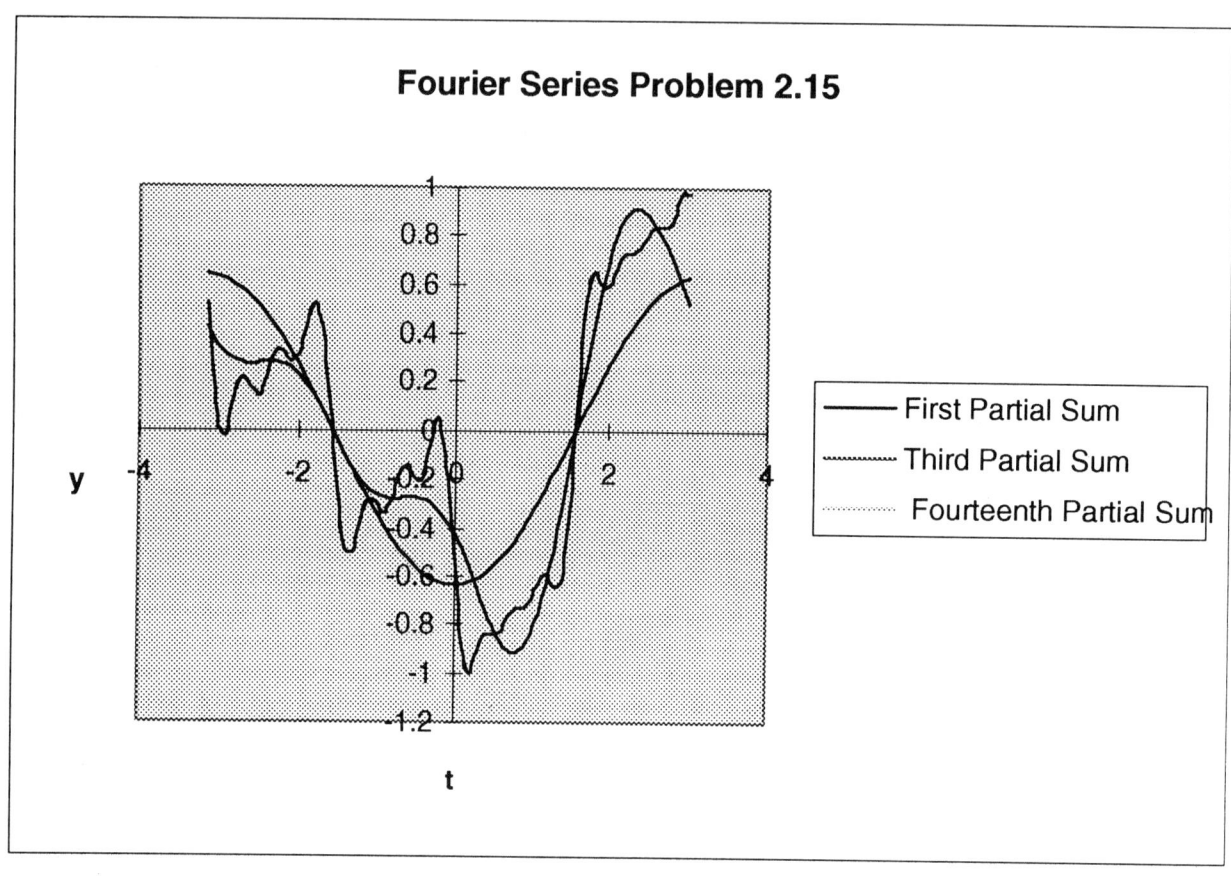

PROBLEM 2.16

KNOWN: $y(t) = t$ for $-5 < t < 5$

FIND: Fourier series for the function $y(t)$.

ASSUMPTIONS: An odd periodic extension is assumed.

SOLUTION:
The function is approximated as shown below

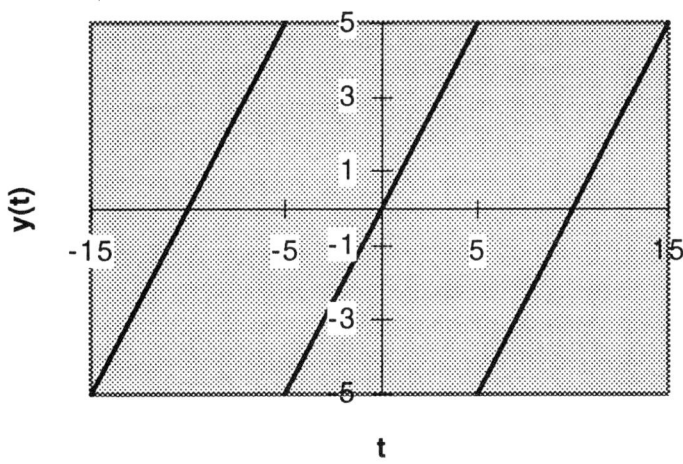

Since the function is odd, the Fourier series will contain only sine terms

$$y(t) = \sum_{n=1}^{\infty} B_n \sin \frac{2n\pi t}{T}$$

where, from (2.17)

$$B_n = \frac{2}{T} \int_{-T/2}^{T/2} y(t) \sin \frac{2n\pi t}{T} dt$$

Thus

$$B_n = \frac{2}{10} \int_{-5}^{5} t \sin \frac{2n\pi t}{10} dt$$

which is of the form x sin ax, and

$$B_n = \frac{2}{10}\left[\left(\frac{5}{n\pi}\right)^2 \sin\frac{2n\pi t}{10} - \frac{10t}{2n\pi}\cos\frac{2n\pi t}{10}\right]_{-5}^{5}$$

$$= \frac{2}{10}\left[\left(\frac{5}{n\pi}\right)^2 \{\sin(n\pi) - \sin(-n\pi)\} - \left(\frac{50}{2n\pi}\right)\{\cos(n\pi) + \cos(-n\pi)\}\right]$$

for n even $B_n = \dfrac{-10}{n\pi}$

for n odd $B_n = \dfrac{10}{n\pi}$

The resulting Fourier series is

$$y(t) = \frac{10}{\pi}\sin\frac{2\pi t}{10} - \frac{10}{2\pi}\sin\frac{4\pi t}{10} + \frac{10}{3\pi}\sin\frac{6\pi t}{10} - \frac{10}{4\pi}\sin\frac{8\pi t}{10} +- \ldots$$

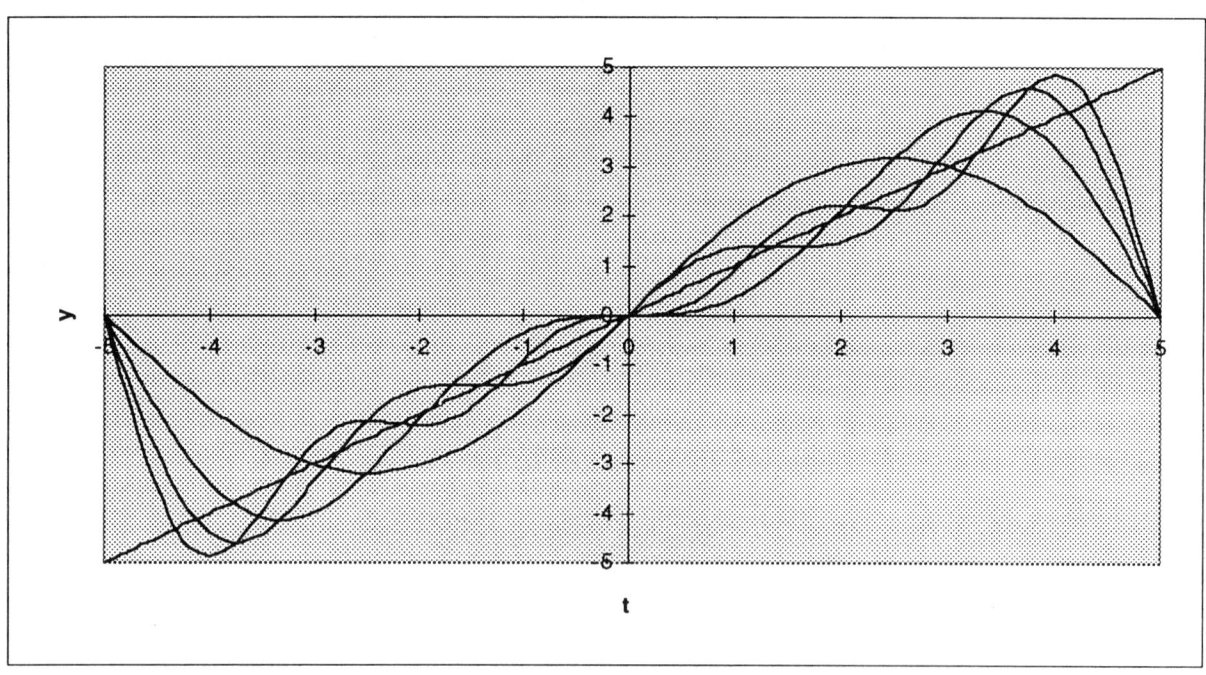

PROBLEM 2.17

KNOWN: $y(t) = t^2$ for $-\pi < t < \pi$; $y(t + 2\pi) = y(t)$

FIND: Fourier series for the function $y(t)$.

SOLUTION:

Since the function $y(t)$ is an even function, the Fourier series will contain only cosine terms,

$$y(t) = A_o + \sum_{n=1}^{\infty} A_n \cos\frac{2n\pi t}{T} = A_o + \sum_{n=1}^{\infty} A_n \cos n\omega t$$

The coefficients are found as

$$A_o = \frac{1}{T}\int_{-T/2}^{T/2} y(t)dt = \frac{1}{2\pi}\int_{-\pi}^{\pi} t^2 dt = \frac{\pi^2}{3}$$

$$A_n = \frac{1}{\pi}\int_{-\pi}^{\pi} t^2 \cos\frac{2n\pi t}{2\pi} dt$$

$$= \frac{1}{\pi}\left[\frac{2t\cos nt}{n^2} + \frac{n^2 t^2 - 2}{n^3}\sin nt\right]_{-\pi}^{\pi}$$

$$= \frac{1}{\pi}\left[\frac{2\pi}{n^2}\cos(n\pi) + \frac{2\pi}{n^2}\cos(-n\pi)\right]$$

for n even $A_n = 4/n^2$ for n odd $A_n = -4/n^2$
and the resulting Fourier series is

$$y(t) = \frac{\pi^2}{3} - 4\left[\cos t - \frac{1}{4}\cos 2t + \frac{1}{9}\cos 3t - + ...\right]$$

a series approximation for π is

$$y(\pi) = \pi^2 = \frac{\pi^2}{3} - 4\left[\cos\pi - \frac{1}{4}\cos 2\pi + \frac{1}{9}\cos 3\pi - + ...\right]$$

or $\quad \frac{\pi^2}{6} = 1 + \frac{1}{4} + \frac{1}{9} + \frac{1}{16} + ...$

PROBLEM 2.18

KNOWN:

$$y(t) = \begin{cases} t & \text{for } 0 < t < 1 \\ 2-t & \text{for } 1 < t < 2 \end{cases}$$

FIND: Fourier series representation of y(t)

ASSUMPTION: Utilize an odd periodic extension of y(t)

SOLUTION:

The function is extended as shown below with a period of 4.

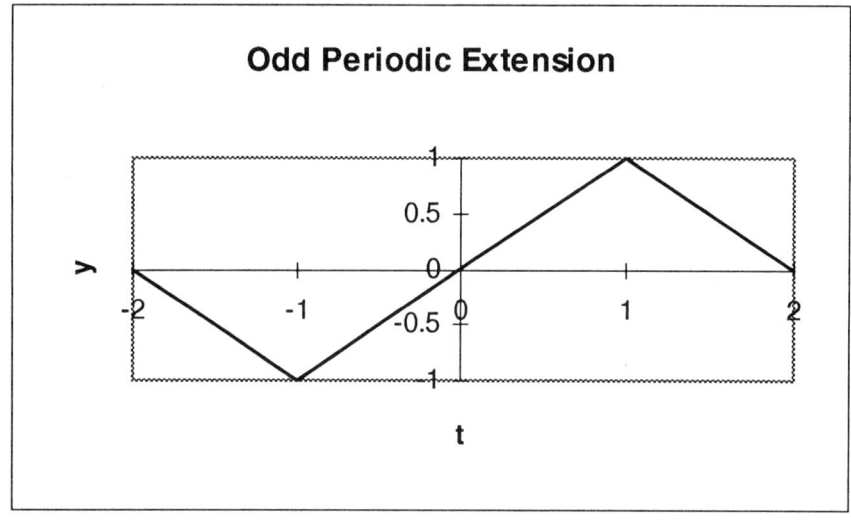

The Fourier series for an odd function contains only sine terms and can be written

$$y(t) = \sum_{n=1}^{\infty} B_n \sin \frac{2n\pi t}{T} = \sum_{n=1}^{\infty} B_n \sin n\omega t$$

where

$$B_n = \int_{-T/2}^{T/2} y(t) \sin \frac{2n\pi t}{T} dt$$

For the odd periodic extension of the function y(t) shown above, this integral can be expressed as the sum of three integrals

$$B_n = \int_{-2}^{-1} -(2+t)\sin\frac{2n\pi t}{4} dt + \int_{-1}^{1} t\sin\frac{2n\pi t}{4} dt + \int_{1}^{2} (2-t)\sin\frac{2n\pi t}{4} dt$$

These integrals can be evaluated and simplified to yield the following expression for B_n

$$B_n = 4\frac{-\sin(n\pi) + 2\sin\left(\frac{n\pi}{2}\right)}{n^2 \pi^2}$$

Since sin(nπ) is identically zero, and sin(nπ/2) is zero for n even, the Fourier series can be written

$$y(t) = \frac{8}{\pi^2}\left[\sin\frac{\pi t}{2} - \frac{1}{9}\sin\frac{3\pi t}{2} + \frac{1}{25}\sin\frac{5\pi t}{2} - \frac{1}{49}\sin\frac{7\pi t}{2} + -\ldots\right]$$

The first four partial sums of this series are shown below

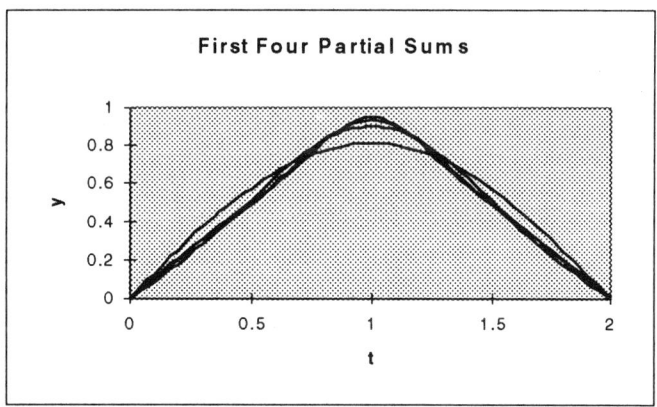

First Four Partial Sums

PROBLEM 2.19

KNOWN:
    a) sin 10t
    b) Figure 2.26

FIND: Classification of signals

SOLUTION:
    a) sin 10t   Deterministic, Simple periodic waveform
    b) Deterministic, Pulse
    c) Deterministic, Periodic (Note: this signal is subject to some variation in interpretation. For example, it may represent a start-up which would eventually reach a steady periodic signal, or it may be interpreted as a complex periodic waveform.)

PROBLEM 2.20

KNOWN: At time zero (t=0)
$$x = 0$$
$$\frac{dx}{dt} = 5 \text{ cm/s} \quad f = 1 \text{ Hz}$$

FIND:
    a) period, T
    b) amplitude, A
    c) displacement as a function of time, x(t)
    d) maximum speed

SOLUTION:

The position of the particle as a function of time may be expressed
$$x(t) = A \sin 2\pi t$$
so that
$$\frac{dx}{dt} = 2A\pi \cos 2\pi t$$

Thus, at t = 0 $\frac{dx}{dt} = 5$

From these expressions we find
a) T = 1 s
b) amplitude, A = 5/2π
c) $x(t) = A \sin 2\pi t$
d) maximum speed = 5 cm/s

PROBLEM 2.21

KNOWN:
    a) Frequency content   c) Magnitude
    b) Amplitude   d) Period

FIND:
Define the terms listed above

SOLUTION:

a) Frequency content - for a complex periodic waveform, refers to the relative amplitude of the terms which comprise the Fourier series for the signal, or the result of a Fourier transform.

b) Amplitude - the range of variation of a particular frequency component in a complex periodic waveform

c) Magnitude - the value of a signal, which may be a function of time

d) Period - the time for a signal to repeat, or the time associated with a particular frequency component in a complex periodic waveform.

PROBLEM 2.22

KNOWN:

Fourier series for the function y(t) = t in Problem 2.16

$$y(t) = \frac{10}{\pi}\sin\frac{2\pi t}{10} - \frac{10}{2\pi}\sin\frac{4\pi t}{10} + \frac{10}{3\pi}\sin\frac{6\pi t}{10} - \frac{10}{4\pi}\sin\frac{8\pi t}{10} + \cdots$$

FIND:

Construct an amplitude spectrum plot for this series.

SOLUTION:

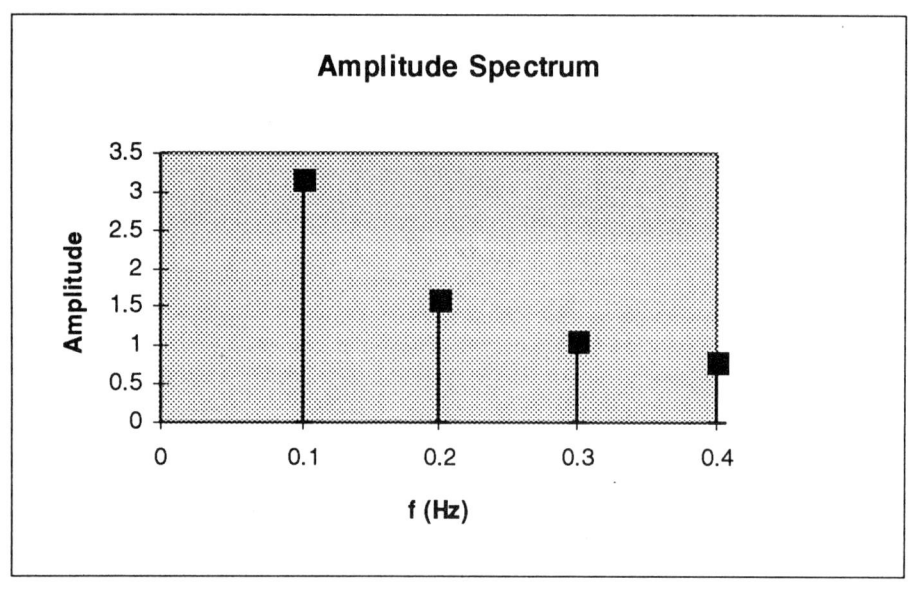

PROBLEM 2.23

KNOWN:
Fourier series for the function $y(t) = t^2$ in Problem 2.17

$$y(t) = \frac{\pi^2}{3} - 4\left[\cos t - \frac{1}{4}\cos 2t + \frac{1}{9}\cos 3t - +...\right]$$

FIND:
Construct an amplitude spectrum plot for this series.

SOLUTION:

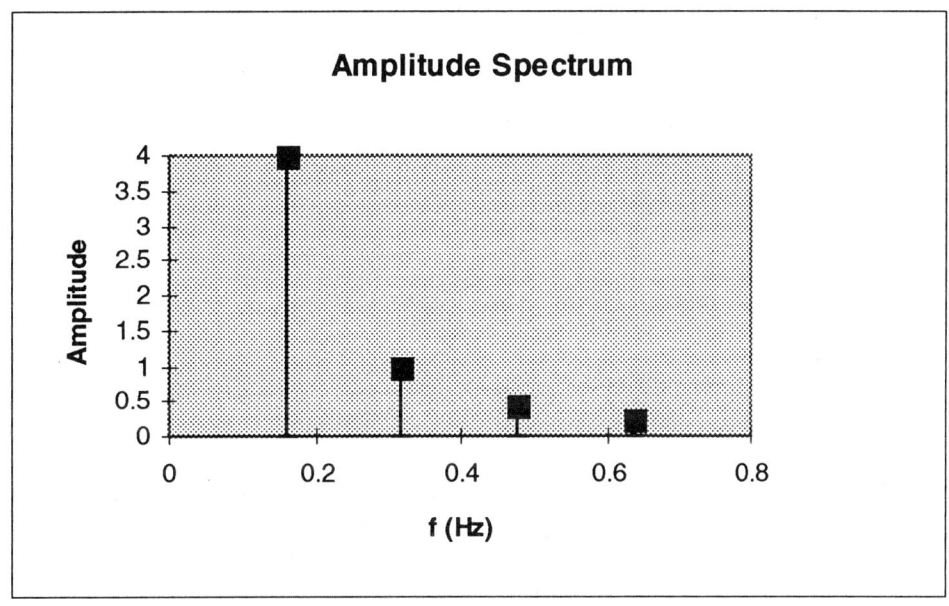

PROBLEM 2.24

KNOWN: Signal sources:

    a) thermostat on a refrigerator

    b) input to a spark plug

    c) input to a cruise control

    d) a pure musical tone

    e) note produced by a guitar string

    f) AM and FM radio signals

FIND: Sketch representative signal waveforms.

SOLUTION: The following signal waveforms are representative of the cases a through f:

a)

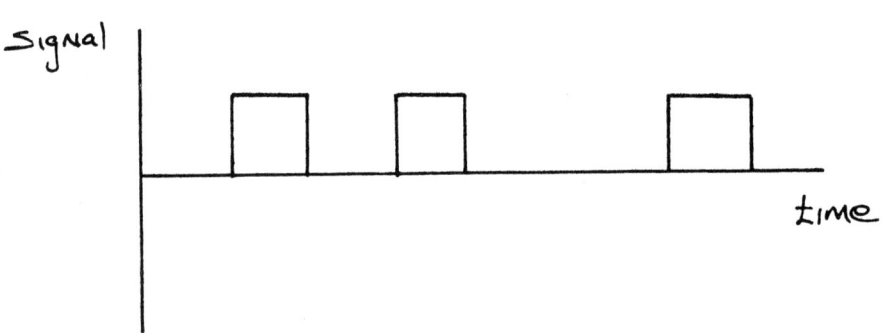

b)

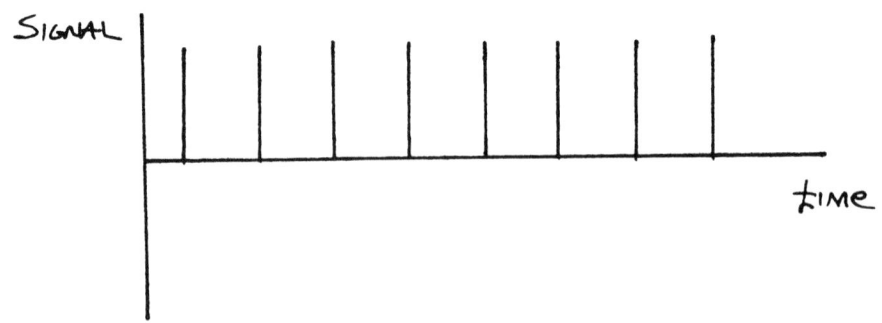

c)

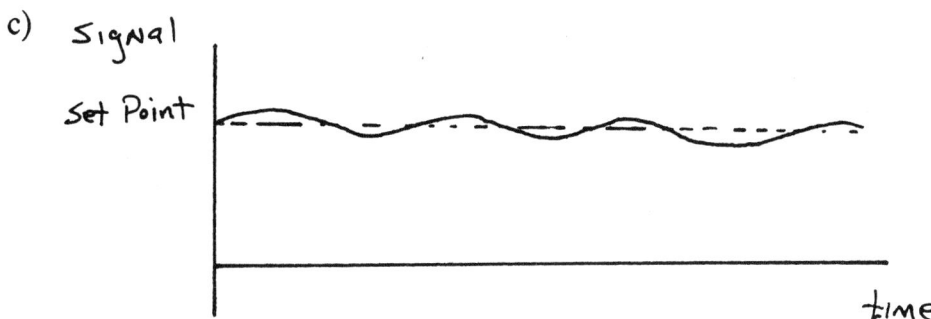

d)

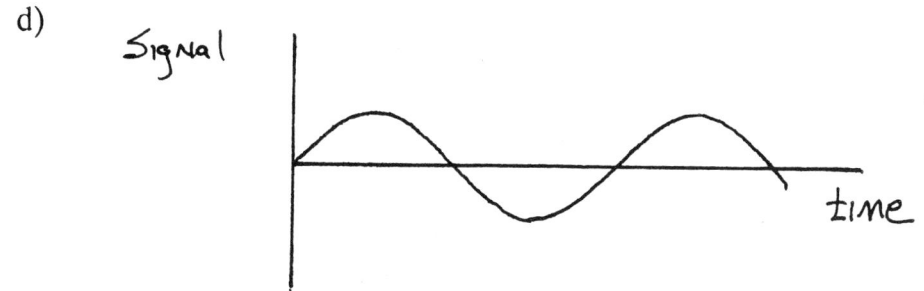

e)

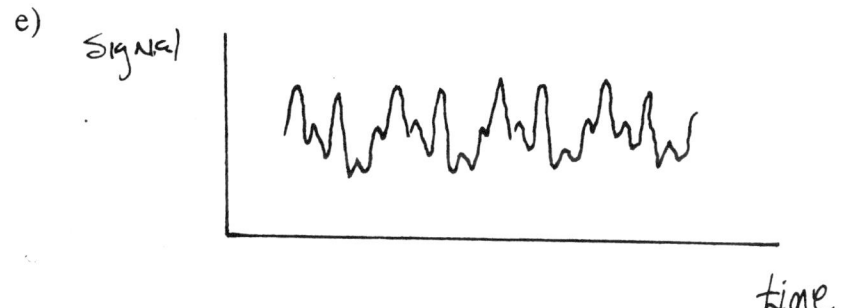

f)

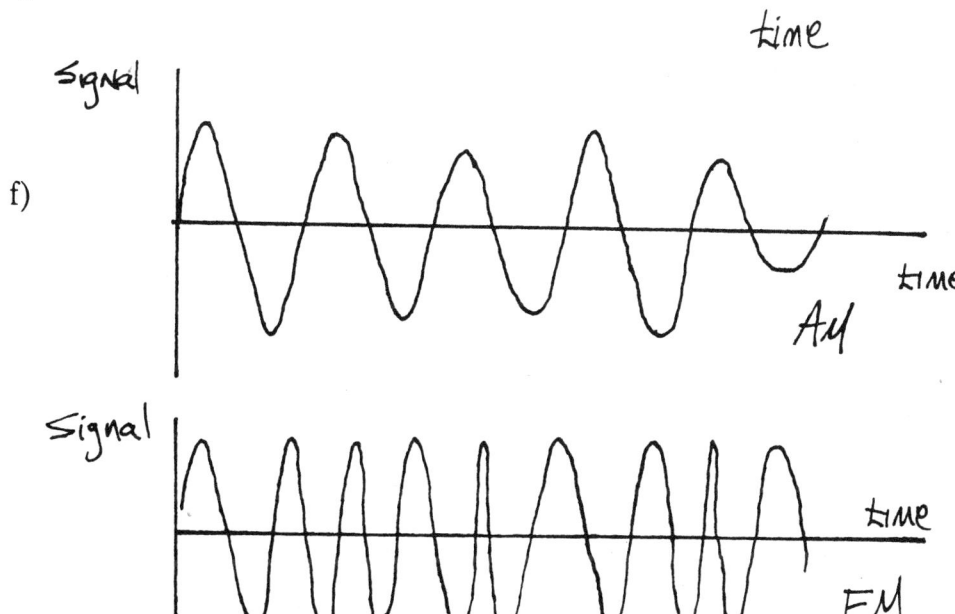

PROBLEM 2.25

KNOWN: $e(t) = 5\sin 31.4t + 2\sin 44t$ volts
$\delta t = 1/N$

FIND: $e(t)$ as a discrete time series of $N = 128$ numbers separated by a time increment of δt. Find the amplitude-frequency spectrum.

SOLUTION

With $N = 128$ and $\delta t = 1/N$, the discrete time series will have a total length of $N\delta t = 1$ second. The signal to be represented contains two fundamental frequencies,

$$f_1 = 31.4/2\pi = 5 \text{ Hz} \quad \text{and} \quad f_2 = 44/2\pi = 7 \text{ Hz}$$

We see that the total time length of the series will represent more than one period of the signal $e(t)$ and, in fact, will represent 5 periods of the f_1 and 7 periods of the f_2 components of this signal. This is important because if we represent the signal by a discrete time series that has an exact integer number of the periods of the fundamental frequencies, then the discrete Fourier series will be exact.

Any DFT or FFT program can be used to solve this problem. Using the program DFT in the Appendix, we can construct the time discrete series and solve for the DFT of the series. This is done by using the function capability [X(L)] of the program to calculate $e(t)$ using a δt = DELT = 1/128 and a total number of N = AN = 2^7 = 128 (i.e. M=7) data points. The program can be written to write the discrete series X(L) directly to a data file or plotting program. The program also returns the amplitude versus frequency information: The time series and the amplitude spectrum are plotted below.

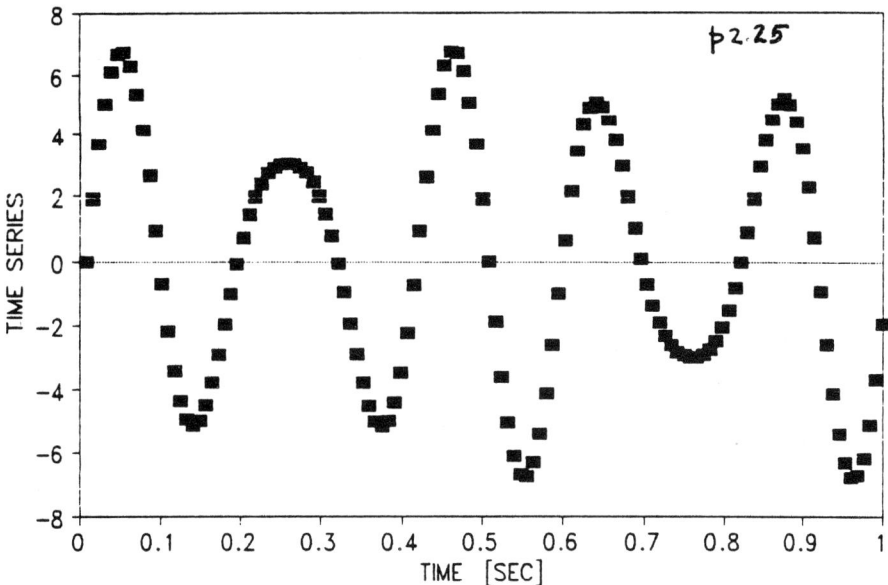

Time Series

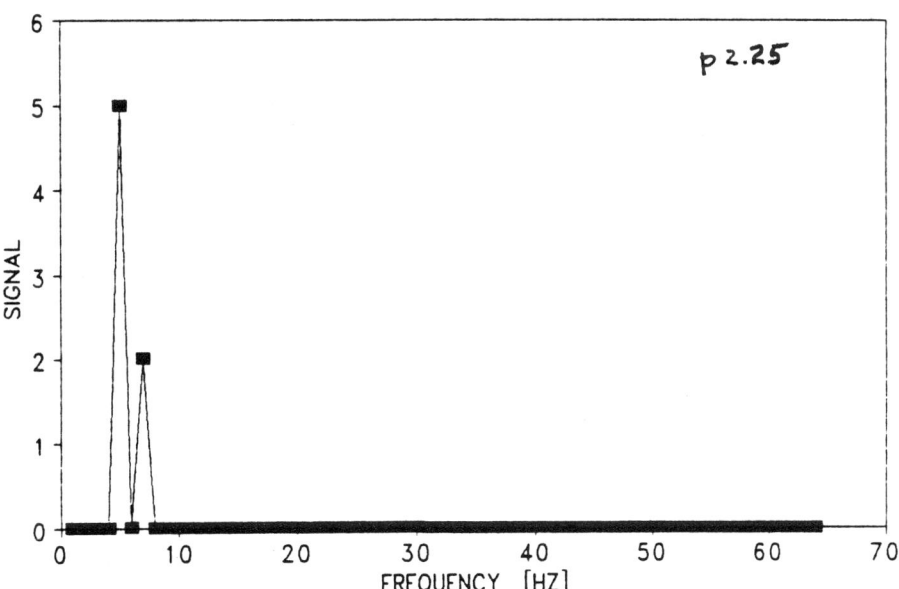

Amplitude Spectrum

PROBLEM 2.26

KNOWN: $e(t) = 5\sin 31.4t + 2\sin 44t$
$\delta t = 1/N$ and $1/2N$
$N = 256$

FIND: $e(t)$ as a discrete time series. The amplitude-frequency spectrum of the time series.

SOLUTION

The signal to be represented contains two fundamental frequencies,

$$f_1 = 31.4/2\pi = 5 \text{ Hz} \quad \text{and} \quad f_2 = 44/2\pi = 7 \text{ Hz}$$

With $N = 256$ and $\delta t = 1/N$, the discrete time series will have a total length of $N\delta t = 1$ second. We see that the total time length of the series will represent more than one period of the signal $e(t)$ and, in fact, will represent 5 periods of the f_1 and 7 periods of the f_2 components of this signal. This is important because if we represent the signal by a discrete time series that has an exact integer number of the periods of the fundamental frequencies, then the discrete Fourier series will be exact. But with $\delta t = 1/2N$, the total time length reduces to $1/2$ second. We can not represent an exact integer number of periods of this signal in $1/2$ second. We should expect that the resulting time series and its DFT will not represent the signal $e(t)$ exactly.

Using the program DFT in the Appendix, we can construct the time discrete series and solve for the DFT of the series. This is done by using the function capability [X(L)] of the program to calculate $e(t)$ using δt = DELT = $1/N$ or $1/2N$ and a total number of N = AN = 2^8 = 256 (i.e. M=8) data points. The program can be written to write the discrete series X(L) directly to a data file or plotting program. The program also returns the amplitude versus frequency information:

$$\text{AMP(K)} = 2. * \text{SQRT}(A(K)**2 + B(K)**2)/\text{AN}$$

where each value of K represents the incremental frequency given by

$$\text{FREQ} = K*\text{RESOL} \quad \text{where RESOL} = 1/N\delta t$$

The time series and the amplitude spectrum are plotted below for both sample time increments.

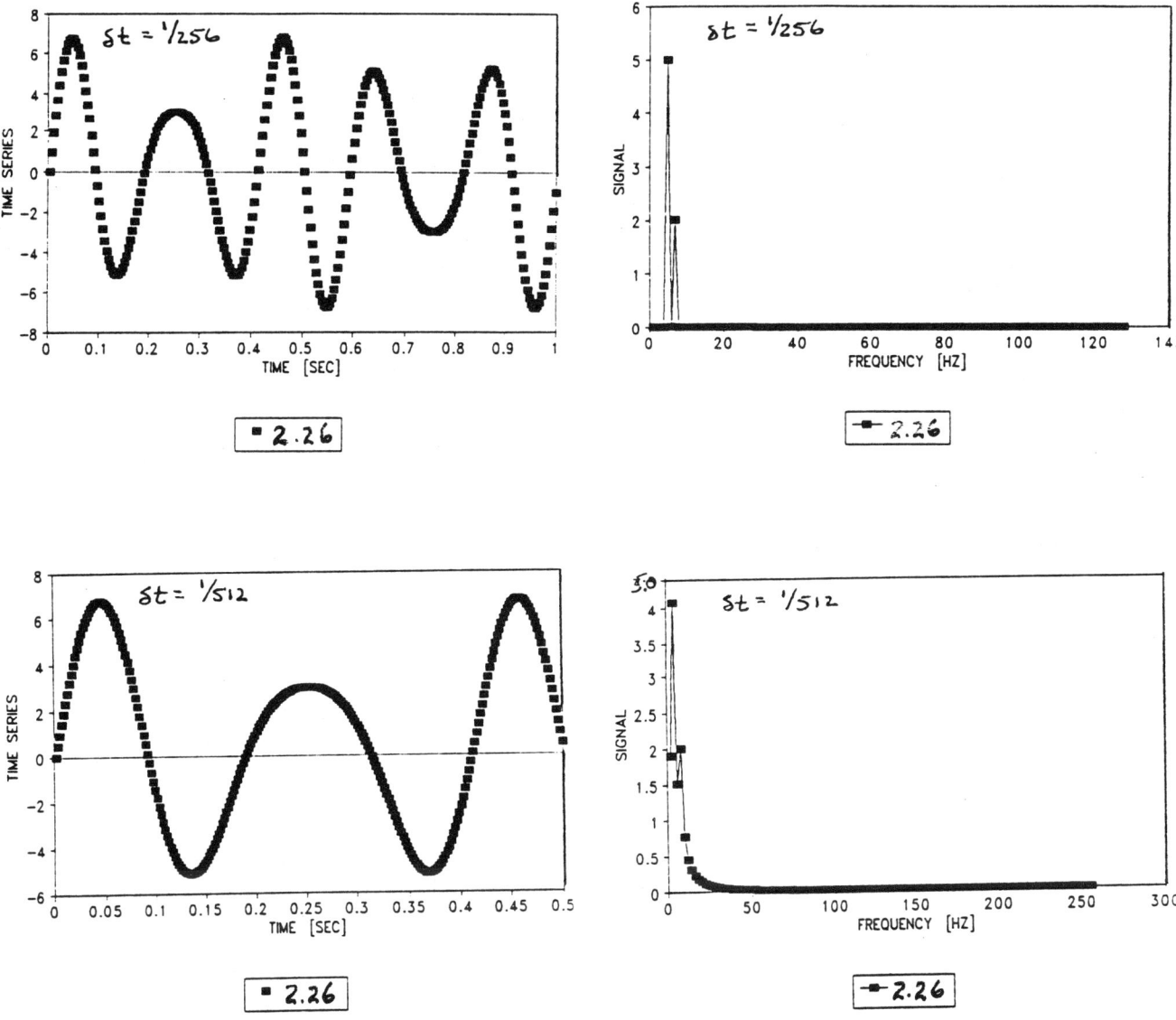

COMMENT

With δt = 1/N, the DFT is exact with amplitude spikes of exactly 5 and 2 at frequencies of 5 and 7 Hz, respectively. It is clear that we can use this information to reconstruct the signal e(t), exactly. However, with δt = 1/2N, the DFT is not exact. While amplitude spikes do occur at 5 Hz and 7 Hz, the amplitude information is smeared ("leaked") about adjacent frequencies. Obviously, using this information to reconstruct e(t) will not yield an exact representation. This effect is called leakage. When the fundamental frequencies are not known, the value of Nδt should be adjusted to minimize leakage.

PROBLEM 2.27

KNOWN: $3250\ \mu s \leq \epsilon \leq 4150\ \mu s$
$f = 1$ Hz

FIND: $y(t)$

SOLUTION

(a) A_o = average value = $(3250 + 4150)/2 = 3700\ \mu s$

(b) $C_1 = (4150 - 3250)/2 = 450\ \mu s$

$f_1 = 1$ Hz

(c) $y(t) = 3700 + 450 \sin(2\pi t \pm \pi/2)\ \mu s$

PROBLEM 2.28

KNOWN: $100 \le F \le 170$ N
$\omega = 10$ rad/s

FIND: $y(t)$

SOLUTION

A_0 = Average value = $(170 + 100)/2 = 135$ N

$C_1 = (170 - 100)/2 = 35$ N

$f_1 = \omega/2\pi = 10/2\pi = 1.59$ Hz

$y(t) = 135 + 35 \sin(10t \pm \pi/2)$

PROBLEM 2.29

KNOWN: $2 \leq x \leq 5$ mm
$f = 100$ Hz

FIND: $y(t)$

SOLUTION

A_o = Average value = $(2 + 5)/2 = 3.5$ mm

$C_1 = (5 - 2)/2 = 1.5$ mm

$f_1 = 100$ Hz

then,

$y(t) = 3.5 + 1.5\sin(200\pi t \pm \pi/2)$ mm

PROBLEM 2.30

KNOWN: The following signals:

a) Clock face having hands
b) Morse code
c) Musical score
d) Flashing neon sign
e) Telephone conversation
f) Fax transmission

FIND: Classify signals as completely as possible.

SOLUTION:

a) Analog, time-dependent, deterministic, periodic, steady-state

b) Digital, time-dependent, nondeterministic

c) Digital, time-dependent, nondeterministic

d) Digital, time-dependent, deterministic

e) Analog or digital, time-dependent, nondeterministic

f) Digital, time-dependent, nondeterministic

PROBLEM 2.31

KNOWN: Amplitude and phase spectrum for $\{y(r\delta t)\}$ from Figure 2.27

FIND: $y(r\delta t)$, δf, δt

SOLUTION

By inspection of Figure 2.27:

$C_1 = 5$ V $C_2 = 0$ V $C_3 = 3$ V $C_4 = 0$ V $C_5 = 1$ V
$f_1 = 1$ Hz $f_2 = 2$ Hz $f_3 = 3$ Hz $f_4 = 4$ Hz $f_5 = 5$ Hz
$\Phi_1 = 0$ r $\Phi_2 = 0$ r $\Phi_3 = 0.2$ r $\Phi_4 = 0$ r $\Phi_5 = 0.1$ r

then, $\delta f = 1$ Hz.

The signal can be reconstructed from the above information:

$$y(r\delta t) = 5\sin(2\pi t) + 3\sin(6\pi t + 0.2) + \sin(10\pi t + 0.1)$$

The exact phase of the signal relative to t = 0 is not known, so $y(r\delta t)$ is true $\pm \pi/2$ in terms of its overall phase.

The DFT returns N/2 values. Therefore, 5 spectral values implies that N = 10. Then:

$$\delta f = 1/N\delta t = 1 \text{ Hz} = 1/(10)\delta t \quad \text{giving} \quad \delta t = 0.1 \text{ s or } f_s = 10 \text{ Hz}$$

Alternatively, by inspection of the plots $f_N = f_s/2 = 5$ Hz giving $f_s = 10$ Hz or $\delta t = 0.1$ s.

PROBLEM 2.32

KNOWN:

$y(t) = (4C/T)t + C$ for $-T/2 < t < 0$

$y(t) = (-4C/T)t + C$ for $0 < t < T/2$

FIND: Show that the signal $y(t)$ can be represented by the Fourier series

$$y(t) = A_o + \sum_{n=1}^{\infty} \frac{4C(1-\cos n\pi)}{(\pi n)^2} \cos \frac{2n\pi t}{T}$$

SOLUTION:

a) Since the function $y(t)$ is an even function, the Fourier series will contain only cosine terms,

$$y(t) = A_o + \sum_{n=1}^{\infty} A_n \cos \frac{2n\pi t}{T} = A_o + \sum_{n=1}^{\infty} A_n \cos n\omega t$$

The value of A_o is determined from Equation (2.17)

$$A_o = \frac{1}{T} \int_{-T/2}^{T/2} y(t) dt$$

$$A_o = \int_{-T/2}^{0} \left(\frac{4Ct}{T} + C\right) dt + \int_{0}^{T/2} \left(\frac{-4Ct}{T} + C\right) dt$$

integrating yields a value of zero for A_o

$$A_o = \frac{1}{T} \left\{ \left[\frac{2Ct^2}{T} + Ct\right]_{-T/2}^{0} + \left[\frac{-2Ct^2}{T} + Ct\right]_{0}^{T/2} \right\} = 0$$

Then to determine A_n

$$A_n = \frac{2}{T} \left\{ \int_{-T/2}^{0} \left(\frac{4Ct}{T} + C\right) \cos \frac{2n\pi t}{T} dt + \int_{0}^{T/2} \left(\frac{-4Ct}{T} + C\right) \cos \frac{2n\pi t}{T} dt \right\}$$

$$A_n = -2 \frac{C(-2 + 2\cos(n\pi) + n\pi \sin(n\pi))}{(n\pi)^2}$$

Since $\sin(n\pi) = 0$, then the Fourier series is

$$y(t) = \sum_{n=1}^{\infty} \frac{4C(1-\cos n\pi)}{(\pi n)^2} \cos \frac{2n\pi t}{T}$$

The values of A_n are zero for n even, and the first three nonzero terms of the Fourier series are

$$\frac{8C}{(\pi)^2}\cos\frac{2\pi t}{T} + \frac{8C}{(3\pi)^2}\cos\frac{6\pi t}{T} + \frac{8C}{(5\pi)^2}\cos\frac{10\pi t}{T}$$

The first term represents the fundamental frequency.

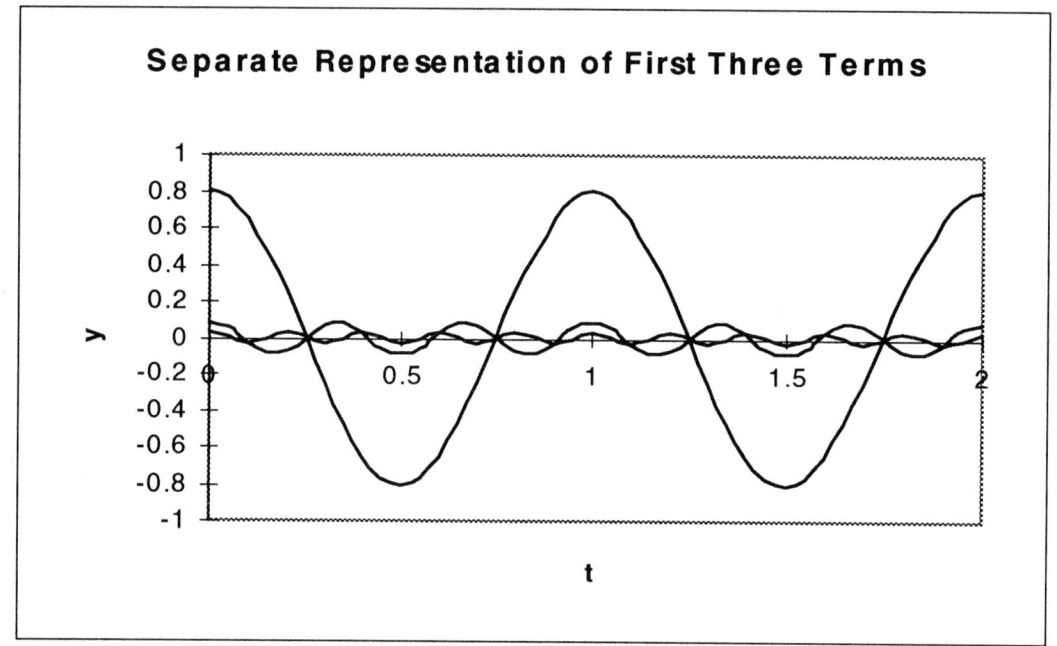

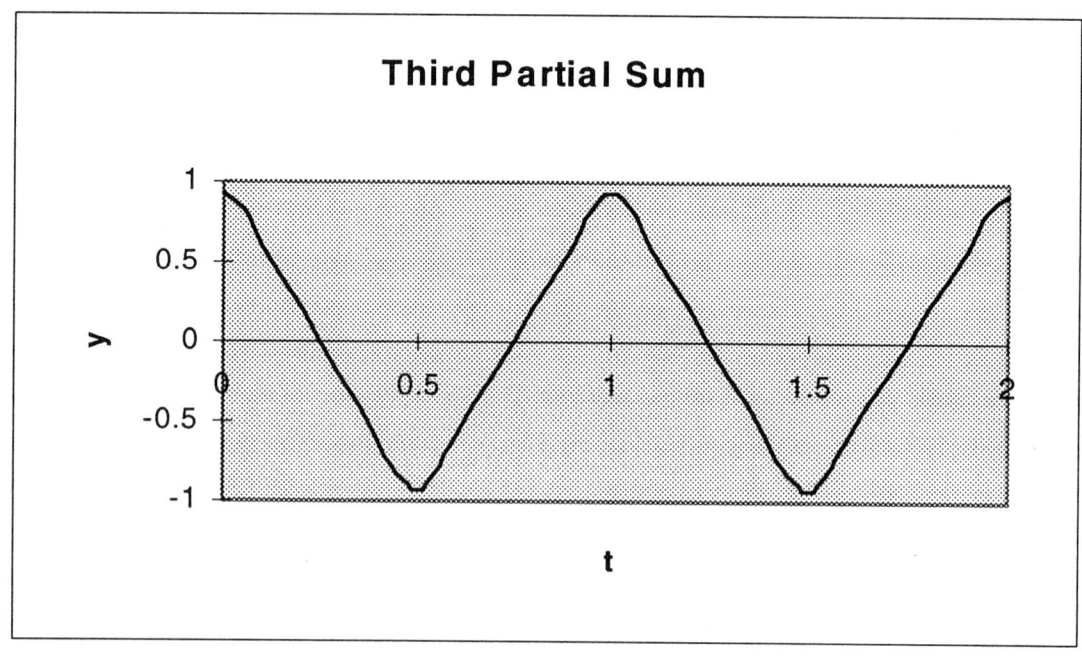

PROBLEM 2.33

KNOWN: Figure 2.16 illustrates the nature of spectral distribution or frequency distribution on a signal.

FIND: Discuss the effects of low amplitude high frequency noise on signals.

SOLUTION:

Assume that Figure 2.16a represents a signal, and that Figures 2.16 b-d represent the effects of noise superimposed on the signal. Several aspects of the effects of noise are apparent. The waveform can be altered significantly by the presence of noise, particularly if rates of change of the signal are important for specific purposes such as control. Generally, high frequency, low amplitude noise will not influence a mean value, and most of the signal statistics are not affected when calculated for a sufficiently long signal.

PROBLEM 3.1

KNOWN: System model

FIND: τ, K, ω_n, ζ as appropriate

SOLUTION

Each of these equations can be fit to either the general form

$$\tau \dot{y} + y = KF(t) \quad \text{first order model}$$

with $\tau = a_1/a_0$, and $K = 1/a_0$, as given by (3.3) and (3.4), or

$$(1/\omega_n^2)\ddot{y} + (2\zeta/\omega_n)\dot{y} + y = KF(t) \quad \text{second order model}$$

with $\omega_n = \sqrt{a_0/a_2}$, $\zeta = a_1/2\sqrt{a_1 a_2}$, $K = 1/a_0$ as given by (3.12) and (3.13). Each of the following solutions is found then by direct comparison.

(a) $10\dot{T} + 5T = 5F(t)$ reduces to $2\dot{T} + T = F(t)$

   Hence, $\tau = 2$ units of time, $K = 1$ unit/unit

(b) $2\ddot{P} + 2\dot{P} + 3P = \sin 4t$ reduces to $0.67\ddot{P} + 0.67\dot{P} + P = 0.33\sin 4t$

   Hence, $\omega_n = 1.22$ r/s, $\zeta = 0.41$, $K = 0.33$ units/unit

(c) $0.15\dot{y} + 14y = 6U(t) = F(t)$ reduces to $0.01\dot{y} + y = 0.07F(t)$

   Hence, $\tau = 0.01$ units of time, $K = 0.07$ units/unit

(d) $10y = 5F(t)$ with $F(t) = 2$ reduces to $y = 0.5F(t)$

   Hence, $K = 0.5$ units/unit

(e) $\ddot{z} + 4\dot{z} + 2z = 1 + 4\sin t = F(t)$ reduces to $0.5\ddot{z} + 2\dot{z} + z = 0.5F(t)$

   Hence, $\omega_n = 1.4$ rad/s, $\zeta = 1.4$, $K = 0.5$ units/unit

(f) $\dot{y} + 10y = 2\sin 6t$ reduces to $0.1\dot{y} + y = 0.2\sin 6t$

   Hence, $\tau = 0.1$ units of time, $K = 0.1$ units/unit

PROBLEM 3.2

KNOWN: System model

FIND: 75%, 90% and 95% response times

ASSUMPTIONS: Unless noted otherwise, all initial conditions are zero.

SOLUTION

We seek the rise time to 75%, 90% and 95% response. For a first order system, the percent response time is found from the time response of the system to a step change in input. The error fraction for such an input is given by

$$\Gamma(t) = e^{-t/\tau}$$

from which the percent response at time t is found by

$$\% \text{ response} = (1 - \Gamma(t)) \times 100$$

$$= (1 - e^{-t/\tau}) \times 100$$

from which t is computed directly. Alternatively, Figure 3.7 could be used.

For a second order system, the system response depends on the damping ratio and natural frequency of the system and can be established from either (3.15) or Figure 3.14.

(a) $0.4\dot{T} + T = 4U(t)$

By direct comparison to the general model of a first order system, $\tau = 0.4$ s. Hence,

$$75\% = (1 - e^{-t/0.4}) \times 100 \quad \text{or} \quad t_{75} = 0.55 \text{ s}$$

$$90\% = (1 - e^{-t/0.4}) \times 100 \quad \text{or} \quad t_{90} = 0.92 \text{ s}$$

$$95\% = (1 - e^{-t/0.4}) \times 100 \quad \text{or} \quad t_{95} = 1.2 \text{ s}$$

Alternatively, Figure 3.7 could be used.

(b) $\ddot{y} + 2\dot{y} + 4y = U(t)$

By direct comparison to equations (3.12) and (3.13),

$\zeta = 0.5$ and $\omega_n = 2$ rad/s

Then using Figure 3.14 as a guide, for a response of

75%: $\omega_n t \simeq 1.75$ so that $t_{75} \simeq 0.9$ s

90%: $\omega_n t \simeq 2$ so that $t_{90} \simeq 1.0$ s

95%: $\omega_n t \simeq 2.4$ so that $t_{95} \simeq 1.2$ s

(c) $2\ddot{P} + 8\dot{P} + 8P = 2U(t)$

By direct comparison to equations (3.12) and (3.13),

$\zeta = 1.0$ and $\omega_n = 2$ rad/s

Then using Figure 3.14 as a guide, for a response of

75%: $\omega_n t \simeq 2.6$ so that $t_{75} \simeq 1.3$ s

90%: $\omega_n t \simeq 3.7$ so that $t_{90} \simeq 1.9$ s

95%: $\omega_n t \simeq 4.6$ so that $t_{95} \simeq 2.3$ s

(d) $5\dot{y} + 5y = U(t)$ reduces to $\dot{y} + y = 0.2U(t)$

By direct comparison to the general model of a first order system, $\tau = 1$ s and $K = 0.2$ units/unit.

$75\% = (1 - e^{-t/1}) \times 100$ or $t_{75} = 1.4$ s

$90\% = (1 - e^{-t/1}) \times 100$ or $t_{90} = 2.3$ s

$95\% = (1 - e^{-t/1}) \times 100$ or $t_{95} = 3.0$ s

PROBLEM 3.3

KNOWN: $K = 4$ V/kg
$F(t) = 10$ kg

FIND: $y(t)$

SOLUTION

Because the input is static and the static output is desired, this system can be modeled by a zero order equation,

$$y = KF = (4 \text{ V/kg})(10 \text{ kg})$$

Hence, $y = 40$ V. It is apparent from the model that if K were to be increased, the static output y would be increased. Note how K takes care of the transfer in the units between input F and output y.

COMMENT

Because we have modeled this system as a zero order responding system, we have eliminated any accommodation for a transient response. This solution for y is valid only under static conditions.

PROBLEM 3.4

KNOWN: let X ≡ % vapor

X	y [units]
0	80
50	40
100	0

FIND: K

SOLUTION

The data fit the linear curve,

$$y = -0.8x + 80$$

The static sensitivity is defined as

$$K = dy/dx \big|_x = -0.80 \text{ units/\% vapor}$$

where K is independent of input x.

COMMENT

Because this system's calibration curve is linear, the static sensitivity remains constant over the input range. Be certain to always provide units for all answers.

PROBLEM 3.5

KNOWN: System model $0.5\dot{y} + y = F(t)$
$K = 1$ unit/unit
$F(t) = 150U(t)$
$y(0) = 100$ units

FIND: $y(t)$

SOLUTION

(a) The solution to the system model was shown to be given by the general form,

$$y(t) = y_\infty + (y(0) - y_\infty)e^{-t/\tau}$$

where here,

$y(0) = 100$ units
$y_\infty = 150$ units
$\tau = 0.5$ s

then,

$$y(t) = 150 + (100 - 150)\, e^{-t/0.5} \text{ units}$$

Alternatively, by direct solution

$$0.5\dot{y} + y = 150U(t)$$

for $t \geq 0^+$

$$y(t) = y_h + y_p$$
$$= Ce^{-t/\tau} + B$$

By substitution, $B = 150$. Then, if $y(0) = 100$ and $\tau = 0.5$, $C = -50$.

$$y(t) = 150 - 50\, e^{-t/0.5}$$

(b) The input signal and output signal are shown below.

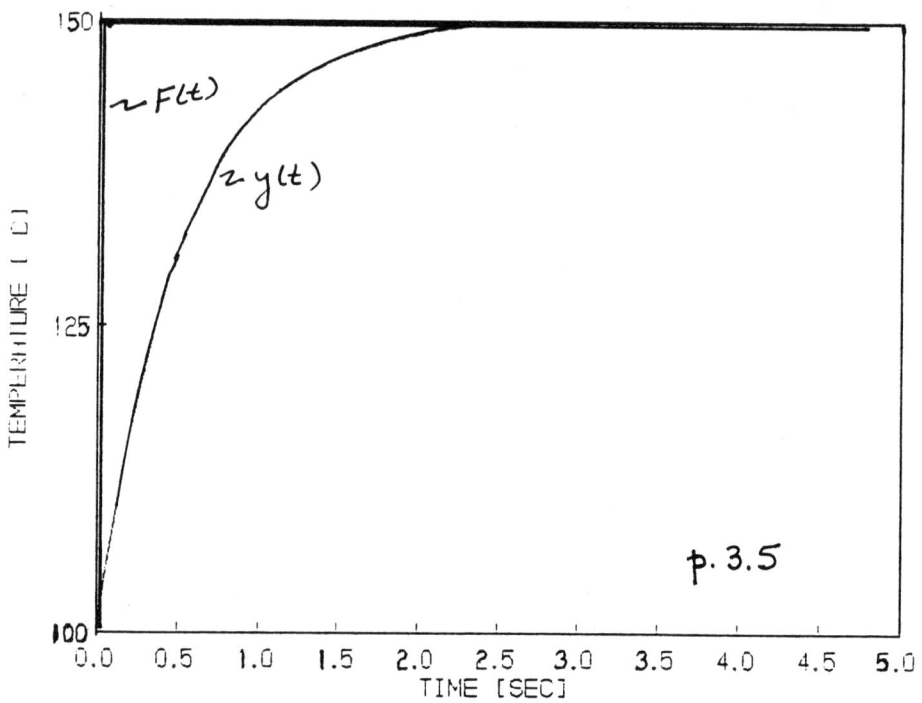

p. 3.5

PROBLEM 3.6

KNOWN: Thermometer similar to Example 3.3
First order system model
$K = 1\ °F/°F$
$\tau = 30\ s$
$F(t) = AU(t) = (120 - 32\ °F)U(t)$
$T(0) = 32\ °F$

FIND: $T(t)$; 90% rise time

SOLUTION

(a) From Example 3.3,

$$mc\dot{T} + hA[T(t) - T(0)] = hA[T_\infty - T(0)]U(t)$$

or

$$\tau\dot{T} + T = 120\ °F \quad \text{for } t \geq 0^+ \text{ with } T(0) = 32\ °F$$

$$30\dot{T} + T = 120\ °F \quad \text{for } t \geq 0^+$$

or

$$T(t) = 120 - 88e^{-t/30} \quad \text{for } t \geq 0^+$$

This response is plotted below.

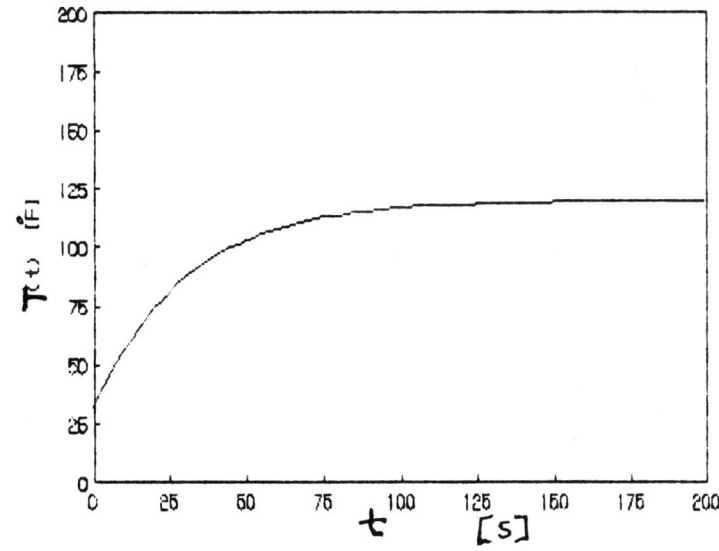

(b) To find the 90% response or rise time,

$$\text{\%response} = 1 - \Gamma(t) = 1 - e^{-t/\tau}$$

which for $\tau = 30$ s yields,

$t_{90} = 69$ s

PROBLEM 3.7

KNOWN: First order instrument
$\tau = 25$ ms

FIND: Rise time

SOLUTION

The percent response of a first order system subjected to a step input is

$$\%\text{response} = 1 - \Gamma(t) = 1 - e^{-t/\tau}$$

For example, for a 90% rise time, $\Gamma = 0.1$. Solving for time,

$$t_{90} = 2.3\tau$$

$$= 5.75 \text{ ms}$$

PROBLEM 3.8

KNOWN: Dynamic calibration using a step input
$y_\infty - y(0) = 100$ units
$y(1.2 \text{ s}) = 80$ units
$y(0) = 0$ units

FIND: $\tau, y(1.5)$

SOLUTION

A first order system subjected to a step input can be modeled as

$$\tau \dot{y} + y = KAU(t) \quad \text{with } y(0)$$

The solution is given by (3.5) as

$$y(t) = KA + (y(0) - y_\infty)e^{-t/\tau}$$

From the information provided, we establish that $KA = 100$ units. Then,

$$y(t) = 100 - 100e^{-t/\tau}$$

Using the information that $y(1.2) = 80$ units, we solve for τ,

$$y(1.2) = 80 \text{ units} = 100 - 100e^{-1.2/\tau} \text{ units}$$

then

$$\tau = 0.75 \text{ s}.$$

At time 1.5 s,

$$y(1.5) = 100 - 100e^{-1.5/0.75} \text{ units} = 86.5 \text{ units}$$

PROBLEM 3.9

KNOWN: First order system
$\tau = 0.7$ s
$KF(t) = KA\sin 4\pi t$

FIND: δ

SOLUTION

The dynamic error is defined as

$$\delta(\omega) = M(\omega) - 1$$

where for a first order system

$$M(\omega) = 1/[1 + (\tau\omega)^2]^{0.5}$$

Now, $\omega \equiv 2\pi f$ so that

$$M(f) = 1/[1 + (2\pi f\tau)^2]^{0.5}$$

By direct substitution,

$$M(f) = 1/[1 + (2.8\pi)^2]^{0.5} = 0.11$$

So that the dynamic error,

$$\delta(2 \text{ Hz}) = \delta(4\pi \text{ rad/s}) = -0.89$$

COMMENT

This result means that the output amplitude of the 2 Hz signal will be 89% smaller than the sensed input amplitude KA. That is, the value KA will be attenuated by a factor of 0.89. Attenuation results with a negative value for δ while a positive value indicates a gain.

PROBLEM 3.10

KNOWN: First order instrument
$\tau = 1$ s
$K = 1$ unit/unit
$F(t) = 10\cos 2.5t = 10\sin(2.5t + \pi/2)$
$y(0) = 0$

FIND: $y_{steady}(t)$, β_1

SOLUTION

For a first order system subjected to a simple periodic waveform input signal, the output response has the form

$$y(t) = ce^{-t/\tau} + M(\omega)KA\sin(2.5t + \pi/2 + \phi(\omega))$$

The steady response is given by the second term on the right side where

$$M(\omega) = 1/[1 + (\omega\tau)^2]^{0.5}$$

$$\phi(\omega) = -\tan^{-1}\omega\tau$$

or these can be found from Figures 3.12 and 3.13. With $\omega = 2.5$ rad/s,

$$M(2.5) = 0.37$$

$$\phi(2.5) = -68^\circ = -1.19 \text{ rad}$$

Then,

$$y_{steady}(t) = (0.37)(1)(10)\sin(2.5t + \pi/2 - 1.19)$$

$$= 3.7\sin(2.5t + 0.38)$$

The time lag arising between input and output signals is given by

$$\beta_1 = \phi(\omega)/\omega$$

$$= -1.19/2.5 = 0.48 \text{ s}$$

COMMENT

The dynamic error in this problem is -63%, a very large number. In effect, the measurement system can not respond quickly enough to follow the input signal. This creates a filtering effect whereby a significant portion of the signal amplitude is attenuated (the term "attenuation" refers to a reduction in value and is indicated by a negative dynamic error). The system is a more effective filter than it is a measuring system. Associated with this large dynamic error is a large phase shift and associated time lag.

PROBLEM 3.11

KNOWN: First order system
$\tau = 2$ s
$0.98 \leq \delta(\omega) \leq 1.02$ required

FIND: The maximum frequency that can be measured ω_{max}

SOLUTION

The dynamic error is defined as

$$\delta(\omega) = M(\omega) - 1$$

A first order system cannot have a value of $M(\omega)$ that is greater than 1. So based on the constraint for dynamic error, we want

$$1 \geq M(\omega) \geq 0.98$$

or

$$0.98 \leq 1/[1 + (\omega\tau)^2]^{0.5}$$

At $\tau = 2$ s, we find

$$\omega \leq 0.10 \text{ rad/s}$$

For this to be true,

$$\omega_{max} = 0.10 \text{ rad/s}$$

or in terms of cyclical frequency

$$f_{max} = 2\pi\omega_{max} = 0.016 \text{ Hz}$$

PROBLEM 3.12

KNOWN: First order system
$\tau = 0.01$ s
$\delta(\omega) \leq \pm 0.10$

FIND: $M(\omega)$, $\phi(\omega)$. Frequency range to meet δ constraint.

ASSUMPTION Input of the form, $F(t) = A \sin \omega t$

SOLUTION

For a first order system subjected to a periodic waveform input, the magnitude ratio and phase shift are given by (3.10) and (3.9) respectively. For this particular system,

$$M(\omega) = 1/[1 + (0.01\omega)^2]^{0.5}$$

$$\phi(\omega) = -\tan^{-1} 0.01\omega$$

or

ω [rad/s]	$M(\omega)$	$\phi(\omega)$
1	1.0	$-0.6°$
10	0.995	$-5.7°$
100	0.707	$-45°$
1000	0.10	$-84°$

Note that at $\omega = \tau$, $M(\omega) = 0.707$, a value that defines the bandwidth of a system.

For a first order system, $M(\omega)$ will always be equal or less than unity. So that the dynamic error constraint reduces to $\delta(\omega) \leq -0.1$. For this to be true,

$$-0.1 \geq (1/[1 + (0.1\omega)^2]^{0.5}) - 1$$

solving,

$$\omega \leq 48.5 \text{ rad/s}$$

It is clear that for all ω such that $0 \leq \omega \leq 48.5$ rad/s the constraint is met.

PROBLEM 3.13

KNOWN: First order system
$\tau = 0.15$ s
$K = 5$ mV/°C
$T(0) = 115$ °C
$F(t) = T(t) = 115 + 12\sin 2t$ °C

ASSUMPTIONS: Output signal is linearly proportional to input signal (that is, K is constant)

FIND: Output response, E(t)
$\delta(\omega), \beta_1$

SOLUTION

We see immediately from the units of static sensitivity that this first order instrument senses temperature and outputs a voltage signal. Hence, a good system model can be written as (3.4),

$$\tau \dot{E} + E = KF(t)$$

where E(t) represents the output voltage signal. Specifically, the system model is written as

$$0.15\dot{E} + E = 5(115 + 12\sin 2t) \text{ mV}$$

with $E(0) = KT(0) = 575$ mV.

To solve for E(t), we assume a solution consisting of both homogeneous and particular parts,

$$E(t) = E_h + E_p$$

Because a first order system has only one root, the solution to the characteristic equation has the form,

$$E_h = Ce^{-t/\tau} = Ce^{-t/0.15}$$

We guess a particular solution of the form,

$$E_p = A + B\sin 2t + D\cos 2t$$

and substitute back into the model with $E(t) = E_p$ and

$$E_p = 2B\cos 2t - 2D\sin 2t$$

This leads to A = 575, B = 55 and D = 16.51. The full solution then has the form,

$$E(t) = Ce^{-t/0.15} + 575 + 55\sin 2t - 16.51\cos 2t$$

Evaluating at E(0) = 575 yields that C = -16.51. Lastly, the measurement system's output response to this input can be rewritten as

$$E(t) = 575 + 57.4\sin(2t - 0.27) - 16.51e^{-t/0.15} \text{ mV}$$

The dynamic error at ω = 2 rad/s can be expressed as

$$\delta(2) = M(2) - 1$$

Using equation 3.10 to solve for $M(\omega)$ yields

$$M(2) = 1/[1 + ((2)(0.15))^2]^{0.5} = 0.96$$

so that $\delta(2) = -0.04$.

The time lag can be found from

$$\beta(\omega) = \phi(\omega)/\omega$$

Using (3.9) at ω = 2 rad/s,

$$\phi(2) = -\tan^{-1}(2)(0.15) = -16.7° = 0.29 \text{ rad}$$

so that the time lag is $\beta(2) = 0.29/2 = 0.15$ s.

PROBLEM 3.14

KNOWN: First order instrument
F(t) = 100U(t) °C
Five seconds are available to interpret signal and provide control signal back to process.

FIND: τ_{max}

SOLUTION

In such a situation, we will want the system to commence a shut-down when the output signal achieves some set threshold value. We must set this threshold value. However, we probably do not wish to set it too low or we run the risk of a unnecessary shut down. Suppose we set the error fraction at 10%, that is at

$$\Gamma \leq 0.10$$

reactor shut down will commence. Then with this value we can find the acceptable time constant using

$$\Gamma = e^{-t/\tau}$$

With Γ = 0.10 and t ≤ 5 s, we solve that $\tau \leq 2.17$ s.

COMMENT

We should note that as the set threshold value is pushed to lower values of error fraction, the value for time constant becomes smaller. For example, at Γ = 0.01, $\tau \leq 0.92$ s. This places a more restrictive design constraint on the sensor and installation selected.

PROBLEM 3.15

KNOWN: Single loop LR circuit (see below) used as filter
$R = 1M$ ohm
$\tau = L/R$

FIND: L such that $M(2000\pi) \le 0.5$

SOLUTION

For the LR filter circuit shown, we can use the voltage loop law to write

$$L\, dI/dt + IR = E_i(t)$$

But across the resistor, $I = E_o(t)/R$. So the system governing equation is given by

$$(L/R)\dot{E}_o + E_o = (1/R) E_i(t)$$

with $E_i(t) = A \sin \omega t$. This is a first order system which has a magnitude ratio defined as

$$M(2000\pi) = 1/[1 + (2000\pi\tau)^2]^{0.5} \le 0.5$$

Solving leads to $\tau \le 0.276$ ms. From which

$$L = R\tau = (1 \times 10^6)(0.000276) = 276\ H$$

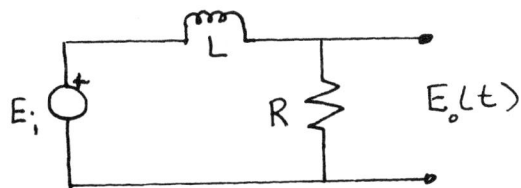

PROBLEM 3.16

KNOWN: $F(t) = 2U(t)$
$y(0) = \dot{y}(0) = 0$

FIND: 90% rise time and settling time

SOLUTION

(a) $8\ddot{y} + 4\dot{y} + 2y = F(t)$

$\omega_n = (2/8)^{0.5} = 1/2$; $\zeta = 4/2[(2)(8)]^{0.5} = 1/2$; $K = 1/2$

From (3.14a), for $t \geq 0^+$

$$y(t) = \tfrac{1}{2}(2) - \tfrac{1}{2}(2)e^{-\tfrac{1}{2}(\tfrac{1}{2})t}\left[\frac{1/2}{\sqrt{1-\tfrac{1}{2}^2}}\sin 0.43t + \cos 0.43t\right]$$

$$= 1 - e^{-t/4}[0.58 \sin 0.43t + \cos 0.43t]$$

where $\omega_d = 0.43$ r/s

By inspection,

90% rise time = 2.74 s

90% settling time = 9.5 s

(b) $3\ddot{y} + 3\dot{y} + 3y = F(t)$

$\omega_n = (3/3)^{0.5} = 1$; $\zeta = 3/2[(3)(3)]^{0.5} = 1/2$; $K = 1/3$

From (3.14a), for $t \geq 0^+$

$$y(t) = \tfrac{2}{3} - \tfrac{2}{3}e^{-t/2}[0.58 \sin 0.86t + \cos 0.86t]$$

where $\omega_d = 0.86$ rad/s

By inspection,

90% rise time = 2.0 s

90% settling time = 4.95 s

(c) $0.9\ddot{y} + 1.2\dot{y} + 0.4y = F(t)$

$\omega_n = (0.4/0.9)^{0.5} = 2/3$; $\zeta = 1.2/2[(0.4)(0.9)]^{0.5} = 1$; $K = 40$

From (3.14b), for $t \geq 0^+$

$y(t) = 80 - 80(1 + 0.67t)e^{-0.67t}$

By inspection,

90% rise time = 5.8 s

90% settling time = 5.8 s (i.e. no ringing)

(d) $0.2\ddot{y} + 1.7\dot{y} + 0.9y = F(t)$

$\omega_n = (0.9/0.2)^{0.5} = 2.12$; $\zeta = 1.7/2[(0.2)(0.9)]^{0.5} = 2$; $K = 1.11$

From (3.14c), for $t \geq 0^+$

$y(t) = 2.22 - 2.22[1.08\, e^{-0.57t} + 0.08\, e^{-7.9t}]$

By inspection,

90% rise time = 4.2 s

90% settling time = 4.2 s (i.e. no ringing)

PROBLEM 3.17

KNOWN: Second order system
$\zeta = 0.6, 0.9, 2.0$

FIND: Plot $M(\omega)$ and $\phi(\omega)$.
Find the frequency range in which $\delta(\omega) \leq \pm 5\%$

SOLUTION

The magnitude ratio and phase shift of a second order system is given by (3.21) and (3.19), respectively,

$$M(\omega) = 1/[(1 - \{\omega/\omega_n\}^2)^2 + (2\zeta\omega/\omega_n)^2]^{0.5}$$

$$\phi(\omega) = \tan^{-1} -(2\zeta\omega/\omega_n)/(1 - [\omega/\omega_n]^2)$$

These are plotted below.

For a constraint of $\delta(\omega) \leq \pm 5\%$, $0.95 \leq M(\omega) \leq 1.05$. This band is indicted on the $M(\omega)$ plots below. Then either from the plots or from (3.21) directly, for $\delta(\omega) \leq \pm 5\%$,

$\zeta = 0.6, \ 0 \leq \omega/\omega_n \leq 0.84$

$\zeta = 0.9, \ 0 \leq \omega/\omega_n \leq 0.28$

$\zeta = 2.0, \ 0 \leq \omega/\omega_n \leq 0.08$

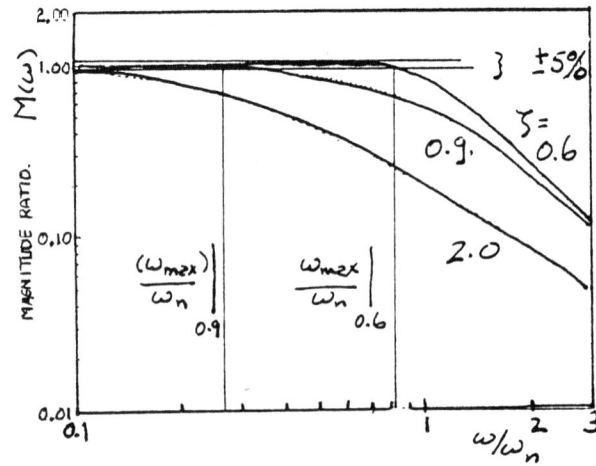

PROBLEM 3.18

KNOWN: Input: $30 \leq C_1 \leq 40°C$ with $f_1 = 10$ Hz
$\delta \leq -0.01$

FIND: $T(t)$, τ_{max}

ASSUMPTIONS: First-order system (i.e. τ requested)
$K = 1$ unit/unit

SOLUTION

The input signal has the form:

$$T(t) = (40 + 30)/2 + [(40 - 30)/2] \sin(2\pi(10Hz)t \pm \pi/2)$$
$$= 35 + 5 \sin(20\pi t \pm \pi/2) \ °C$$

The dynamic error $\delta(\omega) = M(\omega) - 1$ so setting $M(\omega) \geq 0.99$:

$$0.99 \geq 1/[1 + (20\pi\tau)^2]^{1/2}$$

$\tau \leq 2.27$ ms

PROBLEM 3.19

KNOWN: Step test imposed on a second order system
Damped oscillating response, therefore $\zeta < 1$

FIND: Test plan to find ω_n and ζ

SOLUTION

The step test response for an underdamped system is seen in Figure 3.14. The period of oscillation gives $T_d = 2\pi/\omega_d$ where

$$\omega_d = \omega_n(1 - \zeta^2)^{1/2} \qquad (1)$$

and the decay of the oscillations (i.e. the decay of the peaks of each cycle) follows $e^{-\omega_n\zeta t}$.

The test plan should impose the step input and system output recorded at time intervals much less than T_d (e.g. 20 measurements per cycle). The peak values (maximum amplitudes), or alternatively the minimum amplitudes, should be plotted versus time. Because the decay is exponential, that is $y_{max} = e^{-\omega_n\zeta t}$, a plot using semi-log will yield

$$\ln y_{max} = \ln(e^{-\omega_n\zeta t}) = -\omega_n\zeta t$$

and hence the slope of the plot is

$$m = -\omega_n\zeta \qquad (2)$$

Equations 1 and 2 provide for the two unknowns.

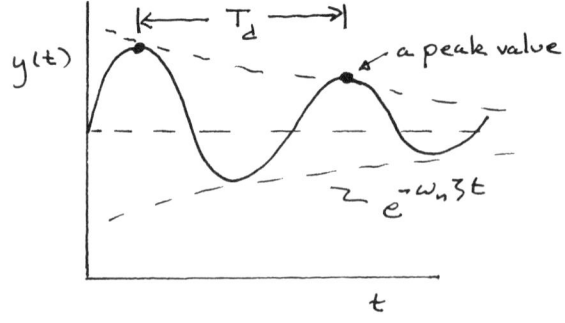
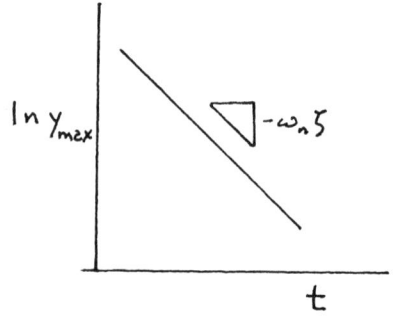

PROBLEM 3.20

KNOWN: $T_d = 0.577$ ms from a step test
second-order system due to oscillations observed

FIND: ω_d

SOLUTION

The period of oscillation of a step test is the period of ringing. Accordingly, the ringing frequency is

$$\omega_d = 2\pi/T_d = 1089 \text{ rad/s}$$

or $f_d = \omega_d/2\pi = 173$ Hz.

PROBLEM 3.21

KNOWN: $\zeta = 0.8$
$\omega_d = 2000\pi$ rad/s (ω [rad/s] $= 2\pi f$ [Hz])
$K = 1.5$ V/V
$F(t) = (12 + 24)/2 + [(24 - 12)/2]\sin 600\pi t = 18 + 6\sin 600\pi t$

FIND: $y_{steady}(t)$

ASSUMPTIONS: Second order system

SOLUTION

The steady response to this input will have the form

$$y_{steady}(t) = KA + KAM(\omega)\sin \omega t + \phi(\omega)$$

where

$$M(\omega) = 1/[(1 - \{\omega/\omega_n\}^2)^2 + (2\zeta\omega/\omega_n)^2]^{0.5}$$

$$\phi(\omega) = \tan^{-1} -(2\zeta\omega/\omega_n)/(1 - [\omega/\omega_n]^2)$$

The natural frequency is readily computed from the measured ringing frequency by

$$\omega_d = \omega_n(1 - \zeta^2)^{0.5}$$

yielding $\omega_n = 3333\pi$ rad/s. Then at $\omega = 600\pi$,

$$M(600\pi) = 0.99$$

$$\phi(600\pi) = -0.289 \text{ rad}$$

Hence,

$$y_{steady}(t) = 27 + 8.9 \sin 600\pi t - 0.289$$

COMMENT

Recall that the steady response is the output signal after all transients have died out. The transient response is found from the homogeneous solution to the equation model.

PROBLEM 3.22

KNOWN: $F(t) = A \sin 500\pi t$
Constraint: $\delta \leq \pm 0.02$
Available $\omega_n = 1200\pi$ rad/s (recall ω [rad/s] $= 2\pi f$ [Hz])
Available values of $0.5 \leq \zeta \leq 1.5$

FIND: ζ required to meet the dynamic error constraint.

ASSUMPTIONS: Second order system

SOLUTION

For $\delta \leq \pm 0.02$, we want $0.98 \leq M(\omega) \leq 1.02$. From the information given, the frequency ratio is $\omega/\omega_n = 0.417$. Then solving for ζ in the equation (3.19),

$$0.98 \leq 1/[(1 - \{\omega/\omega_n\}^2)^2 + (2\zeta\omega/\omega_n)^2]^{0.5}$$

yields

$$\zeta \leq 0.72$$

Repeating at the other limit,

$$1.02 \leq 1/[(1 - \{\omega/\omega_n\}^2)^2 + (2\zeta\omega/\omega_n)^2]^{0.5}$$

yields

$$\zeta \geq 0.63$$

Thus, either a transducer having $\zeta = 0.65$ or 0.7 will work.

PROBLEM 3.23

KNOWN: Second order system responding to a sine input
$2 \le f \le 40$ Hz
$M(f) \le 0.5$

FIND: m, k, c

SOLUTION

Because we wish to attenuate frequency information at and above 2 Hz, we can initiate the design of the system so that it meets the attenuation constraint at 2 Hz. Then we want,

$$M(f) = M(\omega/2\pi) = 1/[(1 - \{\omega/\omega_n\}^2)^2 + (2\zeta\omega/\omega_n)^2]^{0.5} \le 0.5$$

From the freebody diagram below, the system model will have the form,

$$m\ddot{y} + c\dot{y} + ky = F(t)$$

so that, $\omega_n = (k/m)^{0.5}$; $\zeta = c/2(mk)^{0.5}$; $K = 1/k$. Now at $f = 2$ Hz or $\omega = 2\pi f = 12.57$ rad/s, $\omega/\omega_n = 0.4$. Use Figure 3.16 as a guide, note that for $\omega/\omega_n = 0.4$ and $M = 0.5$, we will need to have $\zeta < 2.5$. Also from the Figure, it is apparent that this value will meet the constraint at 40 Hz as well. Suppose we select $m = 2$ kg as a reasonable estimate for the mass of a turntable and select $k = 4000$ N/m, a value that allows the isolation pad to deflect 5 mm. Then, the damping coefficient of the pad should be 447 N-s/m. Of course, the actual values selected will depend on availability.

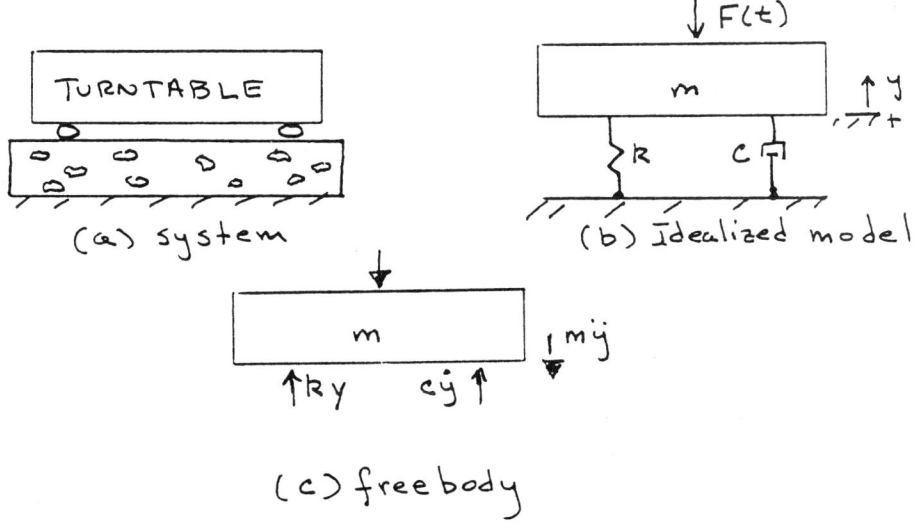

PROBLEM 3.24

KNOWN: RCL circuit: L = 2 H; C = 1 μF; R = 10k Ω
 $E_i(t) = 1 + 0.5 \sin 2000t$ V
 $I(0) = dI(0)/dt = 0$

ASSUMPTIONS: Values for R, C, L remain constant.

FIND: Governing equation and steady output form

SOLUTION

For a single loop RCL circuit, we apply the voltage law to the RCL loop,

$$E_R + E_C + E_L = E_i(t)$$

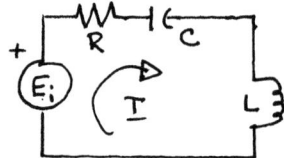

But,

$$E_R = IR; \quad E_L = L\, dI/dt; \quad E_C = (1/C) \int I\, dt$$

Substituting back into the loop equation and differentiating once gives,

$$L\, d^2I/dt^2 + R\, dI/dt + (1/C)\, I = dE_i/dt$$

This is of the form for a second order system. It follows that

$$\omega_n = (1/LR)^{0.5}; \quad \zeta = R/2(LC)^{0.5}; \quad K = 1/C$$

$$\omega_n = 707 \text{ rad/s}; \quad \zeta = 3.54; \quad K = 1 \times 10^{-6}$$

From equations 3.21 and 3.19, respectively with $\omega = 2000$ rad/s,

$$M(2000 \text{ rad/s}) = 0.05$$

$$\phi(2000 \text{ rad/s}) = -1.23 \text{ rad}$$

Then, with $\dot{E}_i(t) = 1000 \sin 2000t$

$$[I(t)]_{steady} = 1 + 0.025 \sin(2000t - 1.23) \; \mu A$$

PROBLEM 3.25

KNOWN: Second order instrument
$\zeta = 0.7$
$\omega_n = 2000\pi$ rad/s
Input: $0 \le \omega \le 1500\pi$ rad/s

FIND: Does transducer meet a ±10% dynamic error constraint?

SOLUTION

The dynamic error is defined by

$$\delta(\omega) = M(\omega) - 1$$

where for a second order system

$$M(\omega) = 1/[(1 - \{\omega/\omega_n\}^2)^2 + (2\zeta\omega/\omega_n)^2]^{0.5}$$

Therefore, the dynamic error constraint can be interpreted as

$$0.90 \le M(0 \le \omega \le 1500\pi \text{ rad/s}) \le 1.10$$

Inspection of Figure 3.16 reveals that the magnitude ratio will not exceed a value of 1.10 over this frequency range. However, its value appears to fall below 0.90. If we test the frequency response at 1500π rad/s, we can check this. Solving with for $M(\omega)$ at $\omega/\omega_n = 0.75$ and $\zeta = 0.7$ yields $M(1500\pi) = 0.88$ for a dynamic error of -12%. So the transducer does not meet the constraint over the entire frequency range of interest.

PROBLEM 3.26

KNOWN: $t_{90} = 100$ ms
$f_d = 1200$ Hz
$\zeta = 0.8$

FIND: $\delta(1\ Hz)$, β_1

SOLUTION

$$f_d = f_n(1 - \zeta^2)^{1/2} \text{ or } f_n = 1200/(1 - 0.8^2)^{1/2} = 2000\ Hz$$

The dynamic error is given by $\delta(f) = M(f) - 1$: $f = 1\ Hz$

$$M(f) = 1/[(1 - \{f/f_n\}^2)^2 + (2\zeta f/f_n)^2]^{0.5}$$

$$= 1/[(1 - \{1/2000\}^2)^2 + (2(.8)(1)/2000)^2]^{0.5} = 1.0$$

so $\delta(1\ Hz) = 0.0$.

The time lag is given by with $f = 1\ Hz$:

$$\beta_1 = \phi/2\pi f = [-\tan^{-1}(2\zeta f/f_n)/\{1 - (f/f_n)\}^2]/2\pi f = 7.3\ ms$$

COMMENT

The near zero dynamic error indicates that the amplitude of the input signal KF(t) and the steady output signal are equal. The time lag indicates that the output signal appears at a time β_1 after the input signal is applied.

PROBLEM 3.27

KNOWN: $\omega_R = 1414$ rad/s
$\zeta = 0.5$
f = 6000 Hz

FIND: $\delta(6000 \text{ Hz})$, $\phi(6000 \text{ Hz})$

SOLUTION

For resonance,

$$\omega_R = \omega_n(1 - 2\zeta^2)^{1/2} \quad \text{or} \quad \omega_n = 1414/(1 - 2(0.5)^2)^{1/2} = 2000 \text{ r/s}$$

or $f_n = 318$ Hz. The ratio $\omega/\omega_n = f/f_n = 6000/318 = 18.87$.

$$M(f) = 1/[(1 - \{f/f_n\}^2)^2 + (2\zeta f/f_n)^2]^{0.5}$$

or $M(6000 \text{ Hz}) = 0.0028$.

The dynamic error,

$$\delta(6000 \text{ Hz}) = M(6000 \text{ Hz}) - 1 = -0.997$$

The phase shift is

$$\phi(6000 \text{ Hz}) = -\tan^{-1} 2(.5)(18.87)/[1 - 18.87^2] = 3.04 \text{ rad} = -174°.$$

So the input signal amplitude is attenuated 99.7% at 6000 Hz with 174° phase shift. The instrument effectively filters out the information at 6000 Hz.

PROBLEM 3.28

KNOWN: Second-order instrument
$0 < \omega \leq 100$ rad/s
Constraint: $\delta(\omega) \leq \pm 10\%$

FIND: Select appropriate values for ω_n and ζ

SOLUTION

The most demanding application will be at the highest input frequency because $\delta = M(\omega) - 1$. Inspection of Figure 3.16 or use of equation 3.21 proves useful to find:

ω_n [rad/s]	ζ	ω/ω_n	$M(\omega)$	$\delta(\omega)$
200	0.4	0.5	1.18	0.18
200	1.0	0.5	0.64	-0.36
200	2.0	0.5	0.47	-0.53
500	0.4	0.2	1.03	0.03
500	1.0	0.2	0.96	-0.04
500	2.0	0.2	0.80	-0.20

So $\zeta = 0.4$ or 1 with $\omega_n = 500$ rad/s will meet the constraint.

PROBLEM 3.29

KNOWN: First order system: $\tau = 0.2$ s; $K = 1$ V/lb
$F(t) = \sin t + 0.3 \sin 20t$ lb (1 lb = 1 pound)

FIND: Steady response output signal

SOLUTION

This system must respond to two frequencies $\omega_1 = 1$ rad/s and $\omega_2 = 20$ rad/s. The steady output will have the form,

$$y_{steady}(t) = KAM(\omega_1)\sin(\omega_1 t + \phi_1) + KAM(\omega_2)\sin(\omega_2 t + \phi_2)$$

For a first order system, we use equations 3.10 and 3.9 to solve for M and ϕ:

$M(1 \text{ rad/s}) = 0.98$ $\phi(1 \text{ rad/s}) = -0.197$ rad

$M(20 \text{ rad/s}) = 0.24$ $\phi(20 \text{ rad/s}) = -1.326$ rad

so that,

$$y_{steady} = 0.98 \sin(t - 0.197) + .072 \sin(20t - 1.326)$$

COMMENT

While the ω_1 information is passed onto the output signal with relatively minor attenuation, the ω_2 information is attenuated by 76%. This measurement system is not a good choice for measuring the ω_2 information.

PROBLEM 3.30

KNOWN: Second order system accelerometer:
$\zeta = 0.5$
$\omega_n = 8000\pi$ rad/s
$\omega = 4000\pi$ rad/s

FIND: $\delta(\omega)$, ω_R

SOLUTION

For any second order system,

$$\delta(\omega) = M(\omega) - 1$$

where $M(\omega)$ is defined by equation 3.21. Then, for this system

$$M(4000\pi) = 1.11$$

so that

$$\delta(4000\pi) = +0.11$$

The output magnitude from this system at this frequency will be 11% greater than the input magnitude, KA. The system fails the criterion of ±10%.

The resonance frequency is found from

$$\omega_R = \omega_n(1 - 2\zeta^2)^{0.5} = 5656\pi \text{ rad/s}$$

PROBLEM 3.31

KNOWN: Second order transducer
$\zeta = 0.4$
$\omega_n = 36000\pi$ rad/s
$F(t) = A \sin 4500\pi t$

FIND: $\delta(\omega), \phi(\omega), \omega_R$

SOLUTION

For any second order system,

$$\delta(\omega) = M(\omega) - 1$$

where $M(\omega)$ is defined by equation 3.21. Then, for this system

$$M(4500\pi) = 1.01$$

so that

$$\delta(4500\pi) = +0.01$$

The phase shift is found from equation 3.19. For this system,

$$\phi(4500\pi) = -0.10 \text{ rad}$$

The resonance frequency is found from

$$\omega_R = \omega_n(1 - 2\zeta^2)^{0.5} = 29686\pi \text{ rad/s}$$

PROBLEM 3.32

KNOWN: Seismic Accelerometer of Example 3.1
$F(t) = x_0 \sin \omega t$

FIND: $M(\omega)$ and $\phi(\omega)$

SOLUTION

From example 3.1,

$$m\ddot{y} + c\dot{y} + ky = c\dot{x} + kx$$

which can be rewritten,

$$\frac{1}{\omega_n^2}\ddot{y} + \left(\frac{2\zeta}{\omega_n}\right)\dot{y} + y = \frac{2\zeta}{\omega_n}\dot{x} + x$$

Assuming zero value initial conditions, $\dot{y}(0) = y(0) = 0$

$$y(t) = y_h + \frac{(\omega/\omega_n^2)\sin(\omega t + \phi)}{\sqrt{(1-(\omega/\omega_n)^2)^2 + (2\zeta\,\omega/\omega_n)^2}}$$

Then,

$$M(\omega) = (\omega/\omega_n)^2 \Big/ \sqrt{(1-(\omega/\omega_n)^2)^2 + (2\zeta\,\omega/\omega_n)^2}$$

$$\phi(\omega) = \tan^{-1}\left[-2\zeta\,\omega/\omega_n \Big/ [1-(\omega/\omega_n)^2]\right]$$

The instrument will be best suited to measure signals having frequency content that is greater than its natural frequency.

COMMENT

The instructor may wish to augment this problem with material from Section 12.5 or refer students to this section of the text for further reading.

PROBLEM 3.33

KNOWN: Second order system pressure transducer
$\zeta = 0.6$
$\omega_n = 8706$ rad/s

FIND: $M(\omega)$, $\phi(\omega)$, ω_R

SOLUTION

For a second order system the magnitude ratio and phase shift are found from equations 3.21 and 3.19, respectively:

$$M(\omega) = 1/[(1 - \{\omega/\omega_n\}^2)^2 + (2\zeta\omega/\omega_n)^2]^{0.5}$$

$$\phi(\omega) = \tan^{-1} -(2\zeta\omega/\omega_n)/(1 - [\omega/\omega_n]^2)$$

Using the known values, we construct the frequency response as

ω [rad/s]	$M(\omega)$	$\phi(\omega)$ [rad]
10	1.0	0
1000	1.0	-0.14
4607	1.04	-0.72
8706	0.83	-1.57
87000	0.01	-3.02

For underdamped systems, the maximum value for $M(\omega)$ occurs at ω_R,

$$\omega_R = \omega_n(1 - 2\zeta^2)^{0.5} = 4607 \text{ rad/s}$$

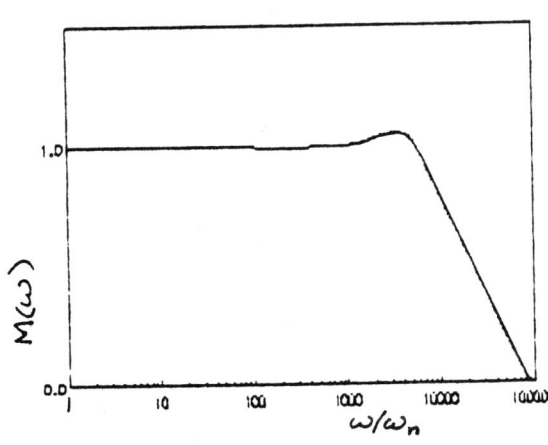
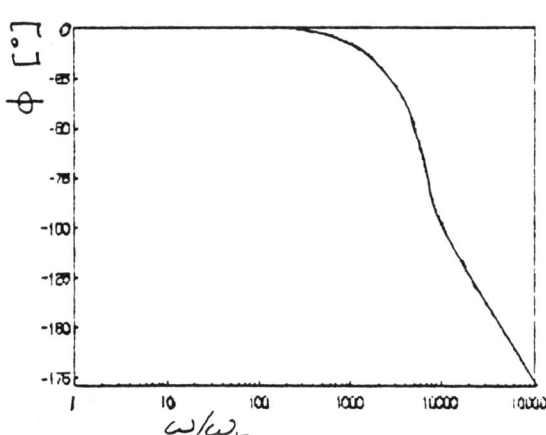

PROBLEM 3.34

KNOWN: Coupled first and second order systems
$F(t) = 10 + 50 \sin 628t$ °C
Input Stage Device:
$\tau = 1.4$ ms
$K = 2$ V/°C
Output Stage Device
$K = 1$ V/V
$\zeta = 0.9$
$\omega_n = 10000\pi$ rad/s

FIND: Steady portion of output signal y(t)

SOLUTION

For coupled systems, call them 1 and 2, the system output is defined by

$$KG(s) = K_1G_1(s)K_2G_2(s) = K_1K_2M_1M_2 e^{i(\phi_1+\phi_2)}$$

with

$$M(\omega)_{system} = M_1(\omega)M_2(\omega)$$

$$K_{system} = K_1K_2$$

$$\phi(\omega)_{system} = \phi_1(\omega) + \phi_2(\omega)$$

For $\omega = 628$ rad/s,

$$M_1 = 1/[1 + (\omega\tau)^2]^{0.5} = 0.75,$$

$$M_2 = 1/[(1 - \{\omega/\omega_n\}^2)^2 + (2\zeta\omega/\omega_n)^2]^{0.5} = 1.0,$$

$$\phi_1 = -\tan^{-1} \omega\tau = -0.721 \text{ rad},$$

$$\phi_2 = \tan^{-1} -(2\zeta\omega/\omega_n)/(1 - [\omega/\omega_n]^2) = -0.036 \text{ rad}$$

Then, with $K_{system} = (2V/°C)(1 V/V)$

$$y_{steady}(t) = 2 + 75 \sin(628t - 0.757) \text{ V}$$

PROBLEM 3.35

KNOWN: Two coupled second order systems
Input signal: $2 \leq C_1 \leq 5$ mm and $f_1 = 85$ Hz
Constraint: $\delta(\omega) \leq \pm 0.05$

FIND: ω_n, ζ for the two measurement system stages

SOLUTION

There are a number of ways to approach this problem and there can not be any one answer. The following is but one approach.

The input function has the form,

$$F(t) = (5+2)/2 + [(5-2)/2]\sin 170\pi t \text{ mm}$$
$$= 3.5 + 1.5 \sin 170\pi t \text{ mm}$$

Now in order to set the specifications, we need to examine how the system as a whole will respond to the input of frequency 170π rad/s. For coupled systems,

$$KG(s) = K_1 G_1(s) K_2 G_2(s) = K_1 K_2 M_1 M_2 e^{i(\phi_1 + \phi_2)}$$

with

$$M(\omega)_{system} = M_1(\omega) M_2(\omega)$$

$$K_{system} = K_1 K_2$$

$$\phi(\omega)_{system} = \phi_1(\omega) + \phi_2(\omega)$$

In general, for second order systems,

$$M(\omega) = 1/[(1 - \{\omega/\omega_n\}^2)^2 + (2\zeta\omega/\omega_n)^2]^{0.5}$$

$$\phi(\omega) = \tan^{-1} -(2\zeta\omega/\omega_n)/(1 - [\omega/\omega_n]^2)$$

The dynamic error of the system is found from

$$\delta(\omega)_{system} = M(\omega)_{system} - 1$$

If we accept, $\delta \leq \pm 0.05$ as a maximum constraint,

$$0.95 \leq M_1 M_2 \leq 1.05$$

We could take, $M_1 = 0.98$ and $M_2 = 1.02$ as one possible design which meets the constraint. With these values selected, we examine each stage in the system.

Input Stage Device

Suppose we fix $\zeta_1 = 0.7$, then for $M_1(170\pi) = 0.98$, $\omega_n = 81\pi$ rad/s. The phase shift then becomes, $\phi_1(170\pi) = -0.71$ rad.

Output Stage Device

Suppose we fix $\zeta_2 = 0.6$, then for $M_2(170\pi) = 1.02$, $\omega_n = 48\pi$ rad/s. The phase shift then becomes, $\phi_2(170\pi) = -0.35$ rad.

COMMENT

The actual values for ζ and ω_n may be limited due to availability from vendors. But the problem demonstrates one approach to dealing with such an open ended problem. Note also that the phase shift for the system is about -0.61 rad, which is not too large for most purposes and is within the range in which phase is linear with frequency.

PROBLEM 3.36

KNOWN: $\omega_R = 82.5$ rad/s
$\zeta = 0.4$
$K = 2$ V/N
$F(t): A_0 = 3$ N, $C_1 = C_2 = 1$ N, $\omega_1 = 8$ rad/s, $\omega_2 = 165$ rad/s

FIND: $y(t)$

SOLUTION

$$y(t) = y_h + 3K + KC_1 M(8) \sin[8t + \phi(8)]$$
$$+ KC_2 M(165) \sin[165t + \phi(165)]$$

For this system the natural frequency is

$$\omega_n = \omega_R/(1 - 2\zeta^2)^{1/2} = (82.5 \text{ r/s})/(1 - 2(0.4)^2)^{1/2} = 100 \text{ rad/s}$$

With $\omega_1/\omega_n = 0.08$ and $\omega_2/\omega_n = 1.65$ and $\zeta = 0.4$, use of Figure 3.16 and 3.17 or equations 3.21 and 3.19 gives:

$M(8 \text{ rad/s}) = 1.004$ (using equations)
$M(165 \text{ rad/s}) = 0.46$
$\phi(8 \text{ rad/s}) = -3.7°$
$\phi(165 \text{ rad/s}) = -142.5°$

The transient response y_h is given by equation 3.14a and appropriate initial conditions.

$$y(t) = y_h + 6 + 2 \sin[8t - 3.7°] + 0.92 \sin[165t - 142.5°]$$

PROBLEM 3.37

KNOWN: First stage: $\tau_1 = 100\ \mu s$, $K_1 = 1\ V/V$
Second stage: $K_2 = 100\ V/V$, $f_n = 15000\ Hz$, $\zeta = 0.8$
Input signal: $F(t) = 5 \sin 2000\ t\ [mV]$

FIND: $y(t)$, $\delta(f)$, β_1

SOLUTION

$$y(t) = y_h(t) + K_1 K_2 M_1 M_2 0.005 \sin[1000t + \phi_1 + \phi_2]\ V$$

For $\omega = 1000\ rad/s$:

$$M_1(1000\ rad/s) = 1/[1 + \{(1000)(100 \times 10^{-3})\}^2]^{1/2} = 0.995$$
$$\phi_1(1000\ rad/s) = -\tan \omega\tau = -5.7°$$

For $\omega/\omega_n = 0.42$ and $\zeta = 0.8$:

$$M_2(1000\ rad/s) = 1/[(1 - \{\omega/\omega_n\}^2)^2 + (2\zeta\omega/\omega_n)^2]^{0.5} = 0.95$$

$$\phi(1000\ rad/s) = \tan^{-1} -(2\zeta\omega/\omega_n)/(1 - [\omega/\omega_n]^2) = -39.2°$$

so: $\beta = \phi/\omega \Rightarrow \beta_1 = 99.4\ \mu s$, $\beta_2 = 684\ \mu s$ and,

$$y(t) = y_h(t) + 0.467 \sin[1000t - 44.9°]\ V$$

COMMENT

The output has been amplified by the second stage. However, a second stage with a higher natural frequency would bring M_2 closer to unity and decrease the phase shift.

PROBLEM 3.38

KNOWN: System spec: $-1d\beta \leq M(0 \leq f \leq 20000 \text{ Hz}) \leq 1d\beta$

FIND: So what?

SOLUTION

This system specification means that for an input frequency within 0 to 20,000 Hz the output amplitude will equal the input amplitude to within $\pm 1d\beta$. From the definition of decibel, $+1d\beta$ is equivalent to $M(f) = 1.12$ or $\delta(f) = 12\%$, while $-1d\beta$ is equivalent to $M(f) = 0.89$ or $\delta(f) = -11\%$. From the specification we can not predict at which frequencies the $\pm 1d\beta$ limits are reached, just somewhere between 0 and 20000 Hz. In fact, a typical audio amplifier will have some spikes and troughs across its frequency response.

Music signals are a series of sinusoidal frequency terms. Even single notes, such as a middle C, consist of a fundamental frequency and harmonics. The harmonics give distinction to the source of the sound so that different instruments are recognizable. Under normal circumstances, we would want the reproduction electronics to neither add nor detract from the signal information (i.e. we want M = 1 across the spectrum).

PROBLEM 3.39

KNOWN: Input Stage Device
$K_1 = 10 \text{ mV/mm}$
$\omega_{n1} = 10000 \text{ rad/s}$
$\zeta_1 = 0.6$
Output Stage Device
$K_2 = 1 \text{ mm/mV}$
$\omega_{n2} = 700 \text{ rad/s}$
$\zeta_2 = 0.7$
$y_{steady}(t) = 90 \sin(4\pi t + \phi_1) + 50\sin(80\pi t + \phi_2)$

FIND: Determine if measurement system specifications are adequate for the input signal.

SOLUTION

Both systems are second order systems.
For the output signal given, the input signal must have the form

$$F(t) = 90/KM(4\pi) \sin 4\pi t + 50/KM(80\pi) \sin 80\pi t$$

For coupled systems,

$$KG(s) = K_1 G_1(s) K_2 G_2(s) = K_1 K_2 M_1 M_2 e^{i(\phi_1 + \phi_2)}$$

Then for second order systems,

$M_1(4\pi) = 1$ $\phi_1(4\pi) = 0$
$M_1(80\pi) = 1$ $\phi_1(80\pi) = 0$
$M_2(4\pi) = 1$ $\phi_2(4\pi) = -0.02 \text{ rad}$
$M_2(80\pi) = 0.99$ $\phi_2(80\pi) = -0.52 \text{ rad}$

Hence,

$M_{system}(4\pi) = 1$ $\phi_{system}(4\pi) = -0.02 \text{ rad}$
$M_{system}(80\pi) = 0.99$ $\phi_{system}(80\pi) = -0.52 \text{ rad}$
$K_{system} = 10 \text{ mm/mm}$

The excellent response characteristics of the measurement system appear to make it a suitable choice for this measured signal.

PROBLEM 3.40

KNOWN: $F(t) = A_1 \sin 170\pi t + A_2 \sin 254\pi t + A_3 \sin 904\pi t$
Input Stage Device Availability
$1000\pi \leq \omega_n \leq 2000\pi$ rad/s
$\zeta = 0.5$
Output Stage Device
$\delta(0.1 \leq f \leq 250k \text{ Hz}) \leq -3$ dB

FIND: Select an acceptable value for ω_n

SOLUTION

There are numerous approaches and a myriad of solutions to this problem. We offer one possible solution. The second order displacement transducer will be most heavily tested at the highest input frequency. Suppose we impose the restriction on the transducer that $\delta_1(\omega) \leq \pm 0.10$ here δ_1 is the dynamic error due to the transducer. Then, for a second order device,

ω_n [rad/s]	$M(904\pi)$	$M(254\pi)$	$M(170\pi)$	ω_R [rad/s]
1000π	1.03	1.03	1.01	707π
1500π	1.14	1.01	1.00	1060π
1750π	1.12	1.01	1.00	1237π
2000π	1.09	1.01	1.00	1414π

Care must be taken in interpreting such numbers. A quick inspection of Figure 3.16 reveals that the input transducer will experience a resonance behavior over these input frequencies. With $\omega_n = 1000\pi$ rad/s, the 904π rad/s input frequency drives the transducer into the post peak resonance region of the graph. It is best to select a transducer where $\omega << \omega_R$. So we select $\omega_n = 2000\pi$ rad/s which should meet this criterion. The spectrum measurement device will not be a factor owing to its wide frequency response relative to these input frequencies.

PROBLEM 4.1

KNOWN: $N > 1000$
$\bar{x} = 9.2$ units
$S_x = 1.1$ units

FIND: Range of x in which 50% of all measurements should fall.

ASSUMPTIONS: Measurand follows a normal density function
Data set sufficiently large such that $S_x \approx \sigma$

SOLUTION

We assume that the data is sufficiently large such that its population behaves as an infinite population. We want to find the interval defined by

$$x' - z_1\sigma \leq x_i \leq x' + z_1\sigma$$

We can find $P(x' + z_1\sigma)$ from the one-sided integral solution to

$$P(z_1) = \frac{1}{\sqrt{2\pi}} \int_0^{z_1} e^{-\beta^2/2} \, d\beta$$

This solution is given in Table 4.3 for $p(z_1) = 0.25$ (one half of the 50% probabilty sought) as $z_1 = 0.674$. Then, we should expect that 50% of the x_i values lie in the interval given by

$$9.2 - 0.7425 \leq x_i \leq 9.2 + 0.7425 \quad (50\%)$$

COMMENT

We can see from Table 4.4 that as N becomes large the value for t approaches a value given by z_1.

PROBLEM 4.2

KNOWN: $N > 10\,000$
$\bar{x} = 204$ units
$S_x = 18$ units

FIND: $x' - z_1\sigma \leq x \leq x' + z_1\sigma$ at $P = 90\%$

ASSUMPTIONS Measurand follows a normal density function
Data set sufficiently large such that $S_x \simeq \sigma$

SOLUTION

Using the definition $z_1 = (x_1 - x')/\sigma$, we find the z_1 value corresponding to the one-sided probability integral $p(z_1) = 0.45$ from Table 4.3. This gives,

$$z_1 = 1.65$$

Then,

$$1.65 = (x_1 - x')/\sigma$$

or

$$1.65\sigma = x_1 - x'$$

or

$$x_1 = x' + 1.65\sigma$$

But a normal (Gaussian) distribution is symmetric about the mean value. Hence, for 90% probability (2)(0.45),

$$x' - 1.65\sigma \leq x \leq x' + 1.65\sigma \quad (90\%)$$

or

$$174.3 \leq x \leq 233.7 \text{ units} \quad (90\%)$$

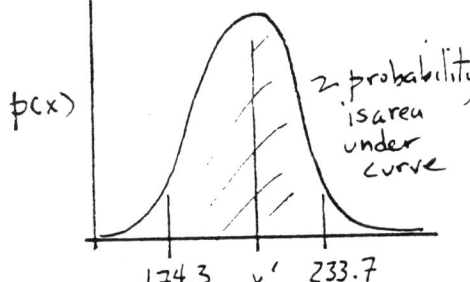

PROBLEM 4.3

KNOWN: $\bar{x} = 121.6$ psi
$S_x = 14$ psi
N is very large

FIND: $P(x > 150 \text{ psi})$

ASSUMPTIONS: Normal distribution
$S_x \simeq \sigma, \bar{x} \simeq x'$

SOLUTION

The z variable is defined by

$$z_1 = (x_1 - x')/\sigma = (121.6 - 150)/14 = -2.028$$

The symmetry of the normal distribution allows that $z_1 = -z_1$ so that we look up $P(2.028)$ from Table 4.3. Interpolation gives

$$P(2.028) = 0.4786$$

This expresses the probability that $121.6 \leq x \leq 150$ psi. Then, the probability that $x > 150$ psi is

$$0.5 - 0.4786 = 0.0214$$

or there is a 2.14% probability that any measurement will yield a value in excess of 150 psi.

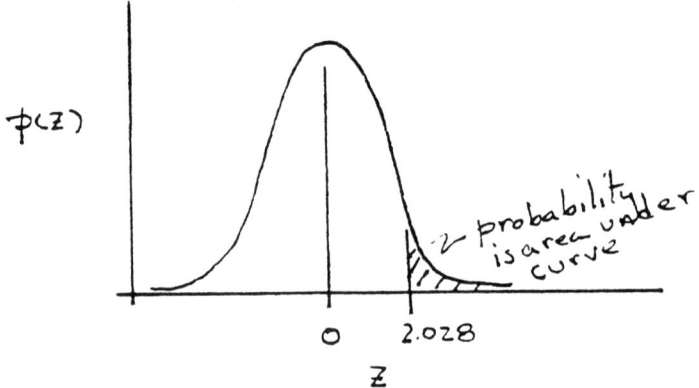

PROBLEM 4.4

KNOWN: Toss of three coins

FIND: Develop the histogram for the outcome of any toss.
State the probability of obtaining three heads on one toss.

SOLUTION

There will be $2^3 = 8$ possible outcomes of any one toss. The probability of three heads is 1 in 8 or 12.5%. The possible outcomes are:

			Number of Heads	n_j
H	H	H		
H	H	T		
H	T	H		
H	T	T	------>	
T	H	H	3	1
T	T	H	2	3
T	H	T	1	3
T	T	T	0	1

The histogram is shown. Because of the few number of tosses, the development of the histogram is primitive. But the symmetry is obvious. As the number of coins is increased, the histogram will become more continuous. This type of distribution is best described as a Binomial distribution (see Table 4.2).

COMMENT

The shape of the binomial distribution will be similar to the Gaussian (normal) distribution except that it will lack the extended "tails" found with the Gaussian shape. The "tails" of the Gaussian distribution are the very low probability extensions to the pdf which accommodate the unlikely but still possible outcomes from random measurements of continuous variables. As the number of possible outcomes (number of coins tossed) becomes large (say 30 or more), the two distributions become nearly identical over a wide interval about the mean and the Gaussian distribution can be used and it is easier. At the extremes of the distribution, the Poisson distribution becomes a good predictor of the binomial probabilities.

PROBLEM 4.5

FIND: Histogram for Table 4.8, Column 1.

SOLUTION

The histogram below is not unique but is only one such solution.

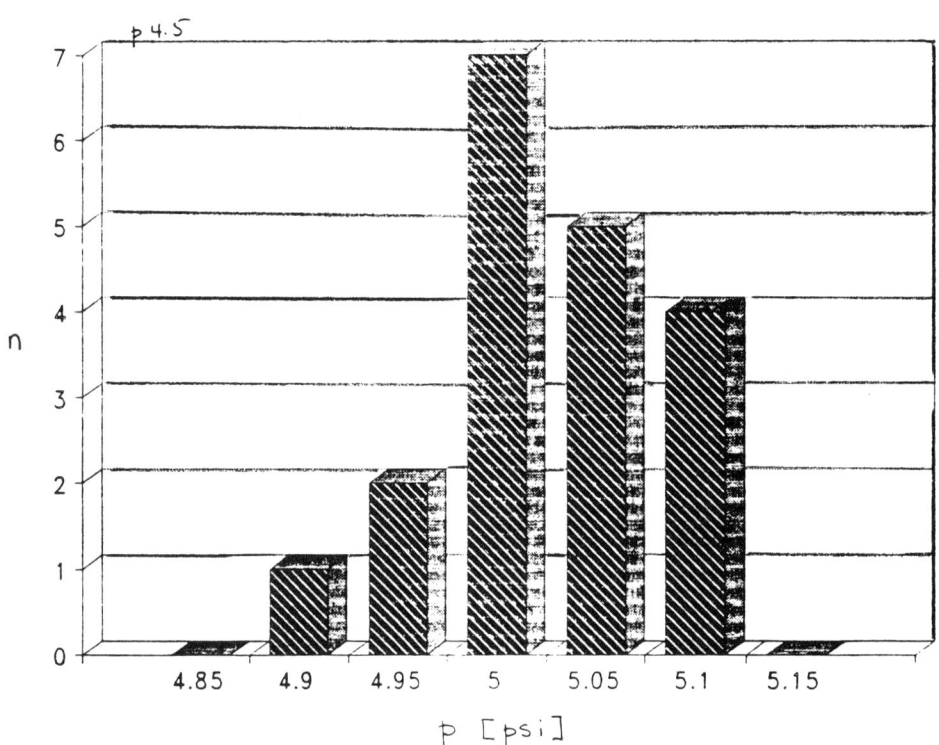

PROBLEM 4.6

FIND: Frequency distribution for Table 4.8, Column 2.

SOLUTION

The distribution below is not unique but is only one such solution.

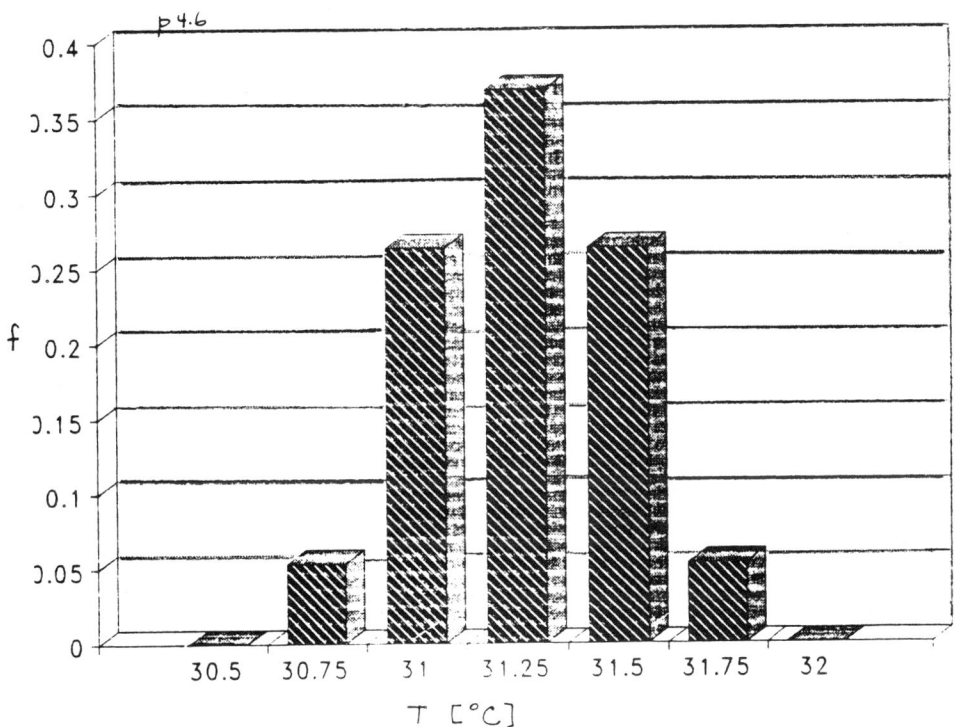

PROBLEM 4.7

FIND: Histogram for Table 4.8, Column 3.

SOLUTION

The histogram below is not unique but is only one such solution.

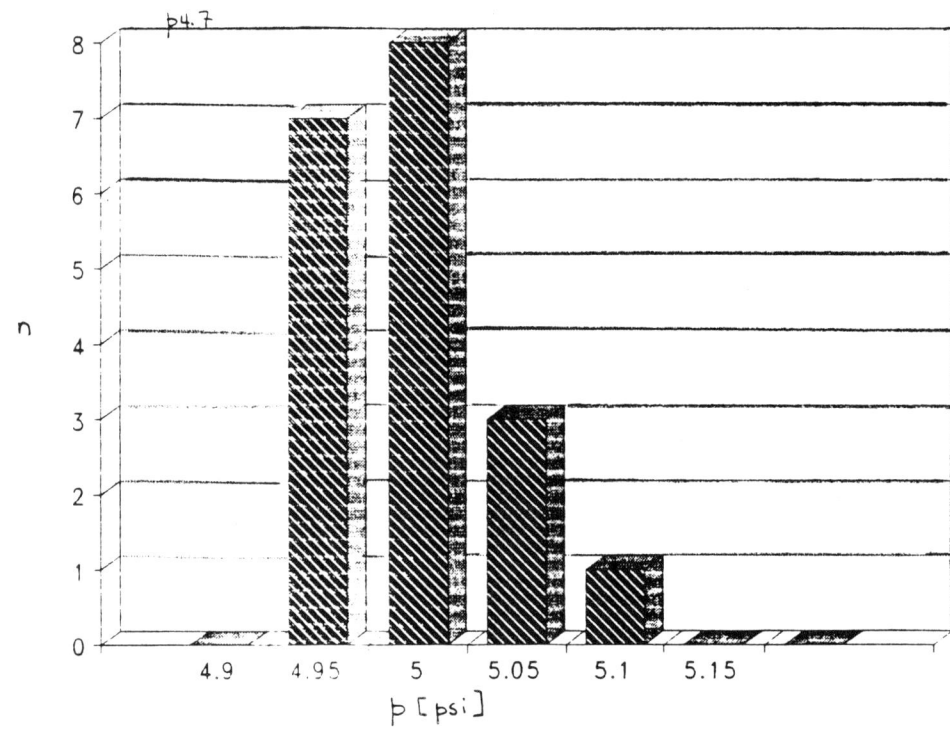

PROBLEM 4.8

KNOWN: Table 4.8
N = 19

FIND: $\bar{p}$, S_p, and ν

SOLUTION

From equation 4.14a, the mean value is

$$\bar{p} = \frac{1}{N}\sum_{i=1}^{N} p_i = 5.01 \text{ kPa}$$

with N = 19. The standard deviation is based on the variance of (4.14b)

$$S_p^2 = \frac{1}{N-1}\sum_{i=1}^{N}(p_i - \bar{p})^2 = 0.003 \text{ kPa}^2$$

such that $S_p = 0.055$ kPa with $\nu = N - 1 = 18$.

PROBLEM 4.9

KNOWN: Data from Table 4.8
  N = 19

FIND: $\bar{p}$, S_p, and ν

SOLUTION

From equation 4.14a, the mean value is

$$\bar{p} = \frac{1}{N}\sum_{i=1}^{N} p_i = 4.97 \text{ kPa}$$

with N = 19. The standard deviation is based on the variance of (4.14b)

$$S_p^2 = \frac{1}{N-1} \sum_{i=1}^{N} (p_i - \bar{p})^2$$

such that S_p = 0.043 kPa. The standard deviation of the means is found (4.16)

$$S_{\bar{p}} = \frac{S_p}{\sqrt{N}} = 0.010 \text{ kPa}$$

with ν = N - 1 = 18.

PROBLEM 4.10

KNOWN: Data of Table 4.8
N = 19

FIND: $\bar{p} \pm tS_p$ (95%)

SOLUTION

From Problem 4.8 (or using equations 4.14a and 4.14b), we found that $\bar{p}$ = 5.01 kPa and S_p = 0.055 kPa with ν = N - 1 = 18. Then from Table 4.4, $t_{18,95}$ = 2.101. From (4.15) the precision of the measured data set can be expressed by

$$p_i = \bar{p} \pm tS_{\nu,P} = 5.01 \pm (2.101)(0.055) = 5.01 \pm 0.116 \; (95\%)$$

here p_i denotes the value of any measurement of pressure, p.

COMMENT

The above probability statement reflects the precision of the data set. In effect, it provides a range of values in which any measured value is expected to fall with 95% probability. A 95% probability implies that 19 out of every 20 measurements are expected to fall in the range defined for p_i.

PROBLEM 4.11

KNOWN: Data of Table 4.8
N = 19

FIND: $\bar{p} \pm tS_{\bar{p}}$ (95%)

SOLUTION

For the 19 data points in Column 1, we find using equations 4.14a and b that $\bar{p}$ = 5.01 kPa and S_p = 0.055 kPa. From equation 4.16, the standard deviation of the means is found to be

$$S_{\bar{p}} = 0.055/19^{0.5} = 0.013 \text{ kPa}$$

with ν = N - 1 = 18. From Table 4.4, $t_{18,95}$ = 2.101. Then, we can expect that the true mean value would lie within the interval defined by (4.17) as

$$p' = \bar{p} \pm tS_{\bar{p}} = 5.01 \pm (2.101)(0.013) = 5.01 \pm 0.027 \text{ kPa} \quad (95\%)$$

A 95% probability implies that 19 out of every 20 complete data sets would show a sample mean value within the range defined for p'. Compare the meaning of this statement to that found in Problem 4.10. They are very different!

COMMENT

The reasoning behind this precision interval for the mean value lies within the limitations of finite statistics. The sample mean value defines the mean value of the 19 data points exactly. But the sample mean is not necessarily the true mean value of the measured variable. However, the true mean value of the measured pressure can be estimated to within some precision interval based on the sample mean and the data set variance. We see that as N → ∞, $S_{\bar{p}}$ → 0, so that $\bar{p}$ → p'. Remember this assumes that there is no bias error acting on the measurement.

PROBLEM 4.12

KNOWN: Data of Table 4.8
N = 19

FIND: $\bar{T} \pm tS_T$ (50%)

SOLUTION

Using equation 4.14a, the mean value for temperature is

$$\bar{T} = \frac{1}{N}\sum_{i=1}^{N} T_i = 31.2°C$$

Based on the value computed for the variance in equation 4.14b, the standard deviation is

$$S_T = \left[\frac{1}{N-1}\sum (T_i - \bar{T})^2\right]^{1/2} = 0.26°C$$

with $\nu = N - 1$. From Table 4.4, $t_{18,50} = 0.688$. Then,

$$T_i = \bar{T} \pm tS_T = 31.2 \pm (0.688)(0.26) = 31.2 \pm 0.18°C \quad (50\%)$$

where T_i is any measured value of temperature, T. Note that the calculations have been rounded to sensible values.

PROBLEM 4.13

KNOWN: Data of Table 4.8
$N = 19$

FIND: $\bar{T} \pm tS_{\bar{T}}$ (50%)

SOLUTION

From Problem 4.12 (or using equations 4.14a and b), we find that $\bar{T} = 31.2°C$ and $S_T = 0.26°C$ with $\nu = N - 1 = 18$. The standard deviation of the means is found using (4.16) as

$$S_{\bar{T}} = 0.26/(19)^{0.5} = 0.06°C$$

From Table 4.4, $t_{18,50} = 0.688$. Then, we can expect that the true mean value would lie within the interval defined by

$$T' = \bar{T} \pm tS_{\bar{T}} = 31.2 \pm (0.688)(0.06) = 31.2 \pm 0.04 \ °C \ (50\%)$$

COMMENT

The reasoning behind this precision interval for the mean value lies within the limitations of finite statistics. The mean value $\bar{T} = 31.2°C$ defines the mean value of the 19 data points exactly but not the true mean value of the measured variable. However, the true mean value of the measured temperature can be estimated from this sample mean value but only to within some precision interval. We see that as $N \to \infty$, $S_{\bar{T}} \to 0$, so that $\bar{T} \to T'$. Remember this assumes that there is no bias error acting on the measurement.

PROBLEM 4.14

KNOWN: Data of Table 4.8
  N = 19

FIND: $\bar{p} \pm tS_p$ (90%)

SOLUTION

From Problem 4.9 (or using equations 4.14a and 4.14b), we found that $\bar{p}$ = 4.97 kPa and S_p = 0.046 kPa with ν = N - 1 = 18. Then from Table 4.4, $t_{18,90}$ = 1.734. Then,

$$p_i = \bar{p} \pm tS_{\nu,P} = 4.97 \pm (1.734)(0.046) = 4.97 \pm 0.08 \quad (90\%)$$

where p_i denotes the value of any measurement of pressure, p.

COMMENT

The above probability statement reflects the precision of the data set. In effect, it provides a range of values in which any measured value is expected to fall with 95% probability. A 90% probability implies that 18 out of every 20 measurements are expected to fall in the range defined for p_i. Note how the interval would change with a change in probability value.

PROBLEM 4.15

KNOWN: Data of Table 4.8
$N = 19$

FIND: $\bar{p} \pm tS_{\bar{p}}$ (90%)

SOLUTION

For the 19 data points in Column 1, we find using equations 4.14a and b that $\bar{p} = 5.01$ kPa and $S_p = 0.055$ kPa. From equation 4.16, the standard deviation of the means is found to be

$$S_{\bar{p}} = 0.055/19^{0.5} = 0.013 \text{ kPa}$$

with $\nu = N - 1 = 18$. From Table 4.4, $t_{18,90} = 1.734$. Then, we can expect that the true mean value would lie within the interval defined by

$$p' = \bar{p} \pm tS_{\bar{p}} = 5.01 \pm (1.734)(0.013) = 5.01 \pm 0.022 \text{ kPa} \quad (90\%)$$

COMMENT

If we compare this answer to that of Problem 4.11, we see that the precision intervals differ. This is due to the different probability levels at which we state the precision intervals. The smaller the probability level, the smaller the precision interval. At 95% probability, we would predict a 1 in 20 chance that the true mean value lies outside the stated interval at 95%. But at 90%, there is a 1 in 10 chance. This is reflected in the smaller interval at 90%.

The reasoning behind this precision interval for the mean value lies within the limitations of finite statistics. The mean value $\bar{p} = 5.01$ kPa defines the mean value of the 19 data points exactly but not necessarily the true mean value of the measured variable. However, the true mean value of the measured pressure can be estimated from this sample mean value but only to within a precision interval. We see that as $N \to \infty$, $S_{\bar{p}} \to 0$, so that $\bar{p} \to p'$. Remember this assumes that there is no bias error acting on the measurement.

Problem 4.14 refers to the precision spread of the data set.

PROBLEM 4.16

KNOWN: Data of Table 4.6, Columns 1 and 3.
  $N = 19$ (repetitions)
  $M = 2$ (replications)

FIND: $\langle \bar{p} \rangle$, $\langle \bar{p} \rangle \pm \langle tS_{\bar{p}} \rangle$ (95%)

ASSUMPTIONS: Data sets in Columns 1 and 3 represent replicate data sets of the same measured variable under similar operating conditions.

SOLUTION

Using equation 4.18, we can find the pooled mean value of the two data sets

$$\langle \bar{p} \rangle = \tfrac{1}{2}(\bar{p}_1 + \bar{p}_2) = 4.99 \text{ psi}$$

and from 4.19a, the pooled standard deviation

$$\langle S_p \rangle = \left[\tfrac{1}{2}(S_{p_1}^2 + S_{p_2}^2)\right]^{1/2} = 0.051 \text{ psi}$$

From (4.20) the pooled standard deviation of the means is

$$\langle S_{\bar{p}} \rangle = \frac{\langle S_p \rangle}{\sqrt{38}} = 0.008 \text{ psi}$$

with degrees of freedom, $\nu = \Sigma (N_j - 1) = 36$. From Table 4.4, $t_{36,95} = 2.028$. Then, the best estimate of the true value is given by

$$p' = \langle \bar{p} \rangle \pm \langle tS_{\bar{p}} \rangle = 4.99 \pm 0.016 \text{ (95\%)}$$

COMMENT

Comparison to Problem 4.16 reveals that the precision interval has been reduced at 95% level due to the increased information found by combining the two data sets.

(over)

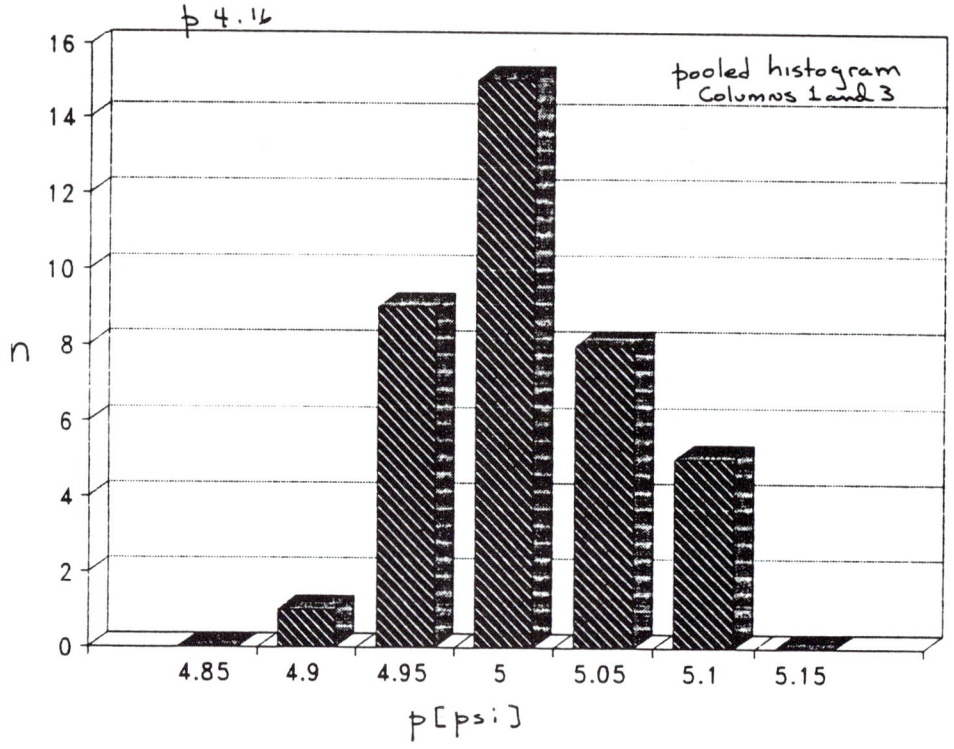

With replication, the larger data set is tending more towards the normal distribution expected.

PROBLEM 4.17

KNOWN: Data of Table 4.6
N = 19

FIND: $\bar{p} \pm tS_p$ (95%)

SOLUTION

From Problem 4.9 (or using equations 4.14a and 4.14b), we found that $\bar{p}$ = 4.97 kPa and S_p = 0.046 kPa with ν = N - 1 = 18. Then from Table 4.4, $t_{18,95}$ = 2.101. Then,

$$p_i = \bar{p} \pm t_{\nu,95} S_p = 4.97 \pm (2.101)(0.046) = 4.97 \pm 0.10 \ (95\%)$$

where p_i denotes the value of any measurement of pressure, p.

PROBLEM 4.18

KNOWN: $x' = 20$ N
$\sigma^2 = 4$ N^2

FIND: $P(23.5 \le x \le 26$ N$)$; $P(16 \le x \le 18.5$ N$)$

SOLUTION

From the definition for $z_1 = (x_1 - x')/\sigma$ and using $\sigma = 2$ N:

(i) $z_a = (23.5 - 20)/2 = 1.75$

$z_b = (26 - 20)/2 = 3$

Then, from Table 4.2, $P(z_a) = 0.4599$ and $P(z_b) = 0.49865$.

$P(23.5 \le x \le 26$ N$) = P(z_b) - P(z_a) = 0.03875$

or a 3.875% chance.

(ii) $z_a = (16 - 20)/2 = 2$

$z_b = (18.5 - 20)/2 = 0.75$

Then, from Table 4.2, $P(z_a) = 0.4772$ and $P(z_b) = 0.2734$.

$P(16 \le x \le 18.5$ N$) = P(z_b) - P(z_a) = 0.2038$

or a 20.38% chance.

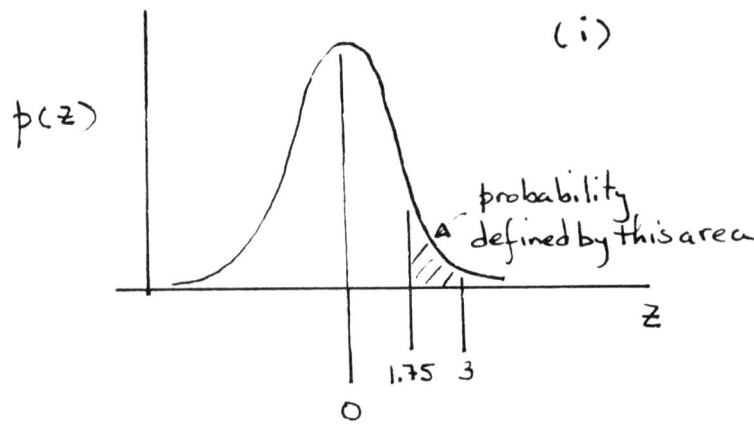

PROBLEM 4.19

KNOWN: Large sample of grades (i.e. infinite statistics applies)

FIND: Number of A, C, D grades awarded

SOLUTION

Referring to the probability graph below:

A: $P(1.6) = 0.4452$ so that the area under the 'A' region is:
$0.5 - P(1.6) = 0.0548$. Hence, 5.48% are A's.

C: $P(0.4) + P(0.4) = 2P(0.4) = 0.3108$
Hence, 31.08% are C's

D: $P(1.6) - P(0.4) = 0.4452 - 0.1554 = 0.2898$
Hence, 28.98% are D's. Likewise, 5.48% are F's and 28.98% are B's.

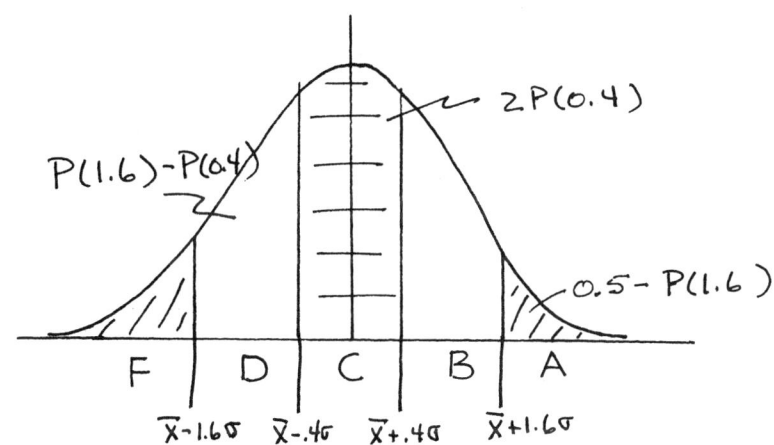

PROBLEM 4.20

KNOWN: p(x) where x is in hours

FIND: mean value of x

SOLUTION

Using (4.4b) the mean value can be found from the known probability density function:

$$\bar{x} = \int_{-\infty}^{\infty} x\, p(x)\, dx = \int_{0}^{\infty} 0.001\, x\, e^{-0.001x}\, dx$$

$$= 1000 \text{ hrs.}$$

The average expected life of the bulb is 1000 hours (60,000 minutes).

PROBLEM 4.21

SOLUTION

The sample mean value estimates the true mean value with a precision indicator given by $S_x/N^{0.5}$. As N increases, the precision improves (i.e. the precision indicator becomes smaller) at a rate of $1/N^{0.5}$.

Hence, the precision in N = 16 measurements improves to only twice that of N = 4 measurements despite quadrupling the number of measurements:

$$S_x/16^{0.5} = S_x/4 \text{ from } S_x/4^{0.5} = S_x/2$$

Likewise, increasing N from 25 to 100 only doubles the precision in the mean value.

COMMENT

For small sample sizes, the gain in precision is impressive for making some extra measurements. But as N increases, this precision gain requires considerably more measurements. Doubling N from 10 to 20 is more efficient than increasing N from 1000 to 2000. This is the "diminishing returns" in using N to improve precision.

PROBLEM 4.22

KNOWN: $N = 270$ with $\bar{x} = 6.92$ MN/m^2 and $S_x^2 = 6.89$ (MN/m^2)^2

FIND: $tS_{\bar{x}}$ at 95%

SOLUTION

The true mean value of these bricks is given by

$$x' = \bar{x} \pm tS_{\bar{x}} \quad (95\%)$$

With $N = 270$, $t_{269,95} \Rightarrow 1.96$, so with $S_x = [6.89 \text{ (MN/m}^2)^2]^{0.5} = 2.62$ MN/m^2

$$\pm tS_{\bar{x}} = \pm (1.96)(2.62 \text{ MN/m}^2)/270^{0.5} = 0.313 \text{ MN/m}^2$$

This is the precision estimate of the true mean value based on the sample mean.

We can state that

$$x' = 6.92 \pm 0.313 \text{ MN/m}^2 \quad (95\%)$$

There is a 95% probability that x' lies between 6.61 and 7.23 MN/m^2.

PROBLEM 4.23

KNOWN: $N = 61$
$\bar{x} = 44.20$ N
$S_x^2 = 4$ N^2

FIND: $P(45.56 \leq x \leq 48.20$ N$)$

SOLUTION

The t value is defined by $t = \pm(x - \bar{x})/S_x$ where $S_x = 2$ N. With this

$$t_a = \pm(45.56 - 44.20)/2 = 0.68$$

Then going to the two sided t chart (Table 4.4) at $\nu = N - 1 = 60$, we find $P \approx 50\%$. So that $P(44.20 \leq x \leq 45.56$ N$) = 0.5/2 = 0.25$. Similarly,

$$t_b = \pm(48.20 - 44.20)/2 = 2.0$$

Again, from the two sided t chart (Table 4.4) at $\nu = N - 1 = 60$, $P \approx 95\%$. So that $P(44.2 \leq x \leq 48.2$ N$) = 0.95/2 = 0.475$. Then,

$$P(45.56 \leq x \leq 48.20 \text{ N}) = 0.475 - 0.25 = 0.225$$

or there is a 22.5% chance that a measured value of x will fall within this interval.

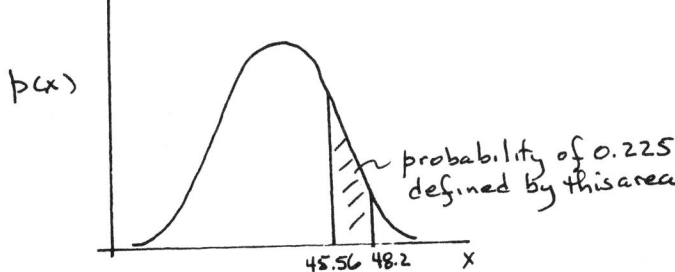

PROBLEM 4.24

KNOWN: M = 3 pooled data sets

FIND: ν, $<\bar{X}>$, $<\bar{X}> \pm <tS_{\bar{X}}>$ (95%)

ASSUMPTIONS: The three data sets represent replicate measurements of a variable under similar conditions.

SOLUTION

For pooled statistics of a single variable with M = 3 replications,

$$\nu = \sum_{j=1}^{M} \nu_j = \sum_{j=1}^{M} (N_j - 1)$$

Then, $\nu = 15 + 20 + 8 = 43$.

From equation 4.22, the weighted pooled mean value is

$$<\bar{X}> = \frac{1}{3}(32 + 30 + 34) = 32 \text{ units}$$

From equation (4.23), the pooled standard deviation is

$$<S_X> = \left[\frac{1}{3}(3^2 + 2^2 + 6^2)\right]^{1/2} = 4.04 \approx 4 \text{ units}$$

From equation 4.24, the standard deviation of the means is

$$<S_{\bar{X}}> = \frac{4}{\sqrt{46}} = 0.6 \text{ units}$$

Then, from Table 4.4, $t_{43,95} \approx 2.018$ (interpolation).

$$X' = <\bar{X}> \pm <tS_{\bar{X}}> \text{ (95\%)} = 32 \pm 1.2 \text{ units (95\%)}$$

COMMENT

In this problem we are faced with three somewhat different results obtained from measuring the same variable on three separate occasions. The variations in the statistics between each data set reflect (1) the ability to duplicate the operating conditions for each test exactly, and (2) the limitations of finite statistics. Please review "replication" discussed in Chapter 1.

PROBLEM 4.25

KNOWN: $\bar{x} = 3027$ psi
$N = 11$
$S_x = 53$ psi

FIND: Does $x' \geq 3000$ psi at 95% probability

SOLUTION

We know that

$$x' = \bar{x} \pm tS_{\bar{x}} \quad (95\%)$$

For $N = 11$, $\nu = 10$ and $t_{10,95} = 2.228$ (Table 4.4). The standard deviation of the means for this data set is

$$S_{\bar{x}} = S_x/N^{0.5} = 16$$

Then,

$$x' = 3027 \pm 35.6 \text{ psi} \quad (95\%)$$

or there is a 95% probability that the true mean strength of the footing is in the interval

$$2991 \leq x' \leq 3063$$

The data suggest the possibility that the footing does not meet the code with a 95% probability. In particular, we know that the footings are prepared in sections as the concrete trucks arrive, unload, and depart. Therefore, portions of the footing almost certainly do not meet the code. Time to break up and repour!

PROBLEM 4.26

KNOWN: Given data set.
$N = 10$

FIND: Check for outliers.

ASSUMPTIONS: Fixed operating conditions.
Measured variable has a normal distribution.

SOLUTION

The statistics for this data set are found to be

$$\bar{x} = \frac{\sum x_i}{N} = 923.7 \text{ N}$$

$$S_x = \left[\frac{1}{N-1} \sum (x_i - \bar{x})^2\right]^{1/2} = 8.13 \text{ N}$$

The modified three sigma test introduces the modified z variable, z_0

$$z_0 = |(x_i - \bar{x})/S_x|$$

A survey of the data indicates that data point #3 could be suspect. Computing a value for z_0 with $x_i = 908$ N gives, $z_0 = 1.93$. From Table 4.3, $P(1.93) = 0.4732$. Then,

$$N[0.5 - P(1.93)] = 0.27$$

Since this value exceeds 0.1, the value for data point #3 apparently falls within the bounds to be expected from normal scatter and is NOT an outlier. No outliers are detected in this data set.

Using $t_{9,95} = 2.262$, and $S_{\bar{x}} = S_x/N^{0.5} = 2.57$,

$$x' = 923.7 \pm 5.8 \text{ N} \quad (95\%)$$

PROBLEM 4.27

KNOWN: N = 20 measurements taken from a large batch
Sample statistics: $\bar{x}$ = 47.5 mm S_x = 8.4 mm
Claim: x' = 42.1 mm (for batch)

FIND: Is claim supported by the data set?

ASSUMPTION: Sample is representative of the batch.

SOLUTION

For N = 20, ν = 19 so that at the 95% level, $t_{19,95}$ = 2.093. The true mean based on the batch statistics is

$$x' = 47.5 \pm (2.093)(8.4)/20^{0.5} = 47.5 \pm 3.93 \text{ mm} \quad (95\%)$$

The claim is not supported by the sample.

PROBLEM 4.28

KNOWN: Data set provided.
N = 5

FIND: Best curve fit to the data set and 95% confidence interval.

SOLUTION

The data can be fit to: $y_c = a_0 + a_1 x + ... + a_m x^m$

m	a_0	a_1	a_2	a_3	ν	$t_{\nu,95}$	S_{yx}	tS_{yx}
1	1.90	1.37	----	----	3	3.182	0.648	2.06
2	2.70	0.44	0.17	----	2	4.303	0.37	1.59
3	1.89	2.25	-0.71	0.11	1	12.706	0.09	1.14

A third order fit reduces S_{yx} to a minimum. The data are plotted below.

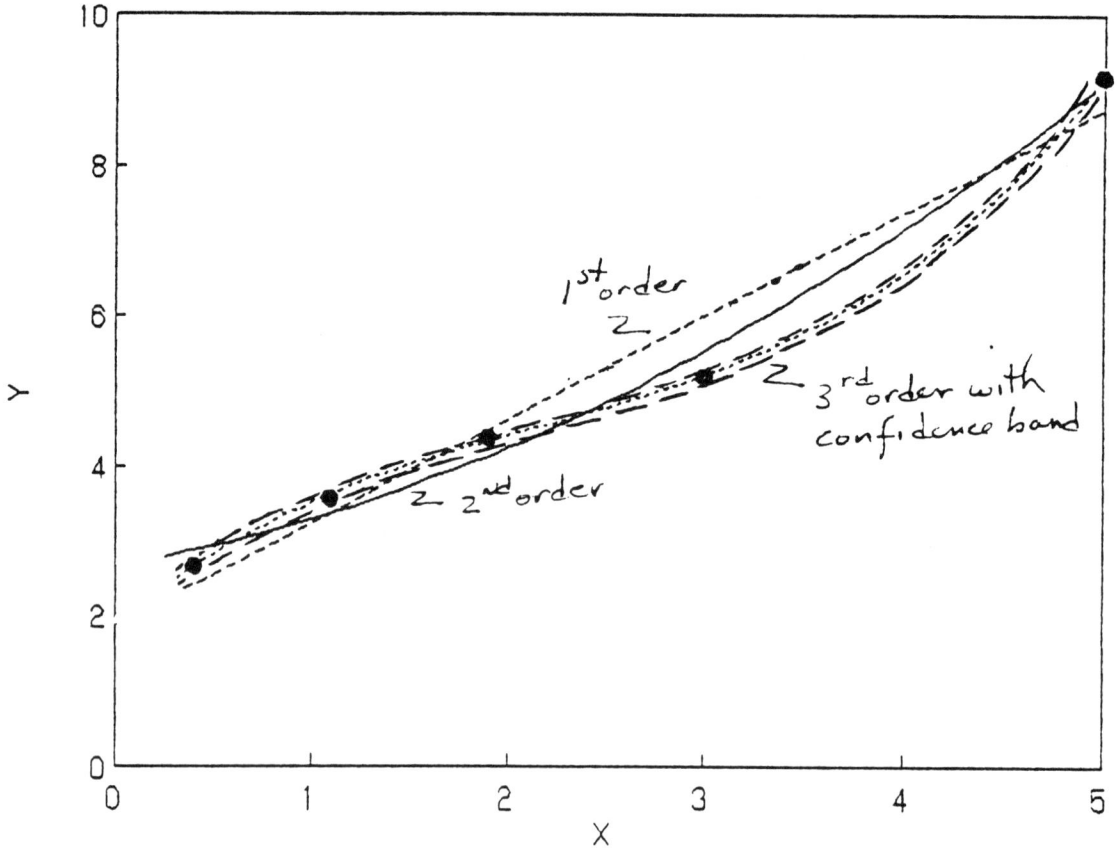

PROBLEM 4.29

KNOWN: Data set provided.
N = 6

FIND: Best curve fit to the data set and 95% confidence interval.

SOLUTION

The data are plotted below and fit to the curve $y_c = 5.05x^{2.3}$. The t value is found to be $t_{4,95} = 2.770$ and $S_{yx} = 0.025$ on the log-log plot.

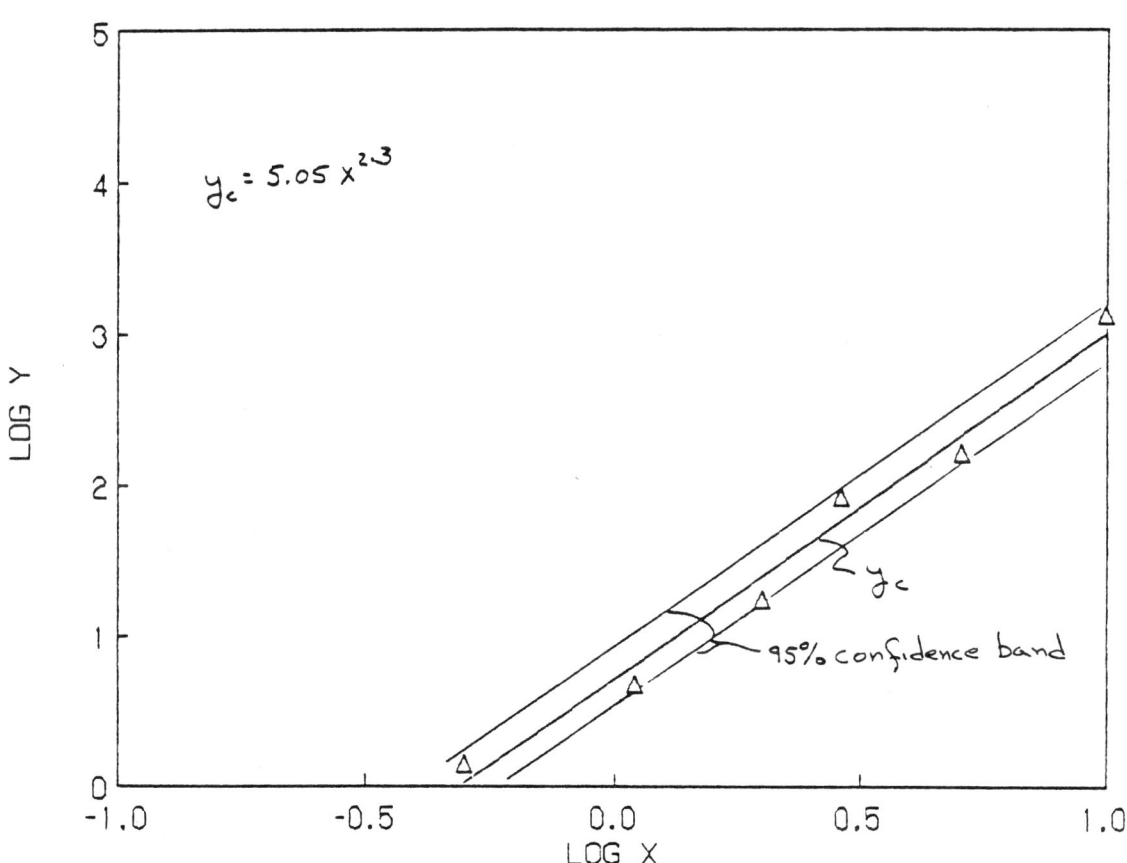

PROBLEM 4.30

KNOWN: Data set provided.
N = 4

FIND: Best curve fit to the data set and 95% confidence interval.

SOLUTION

The data are plotted below and fit to the curve $y_c = 0.58x^{2.03}$. The t value is found to be $t_{2,95} = 4.303$ and $S_{yx} = 0.021$ on the log-log plot.

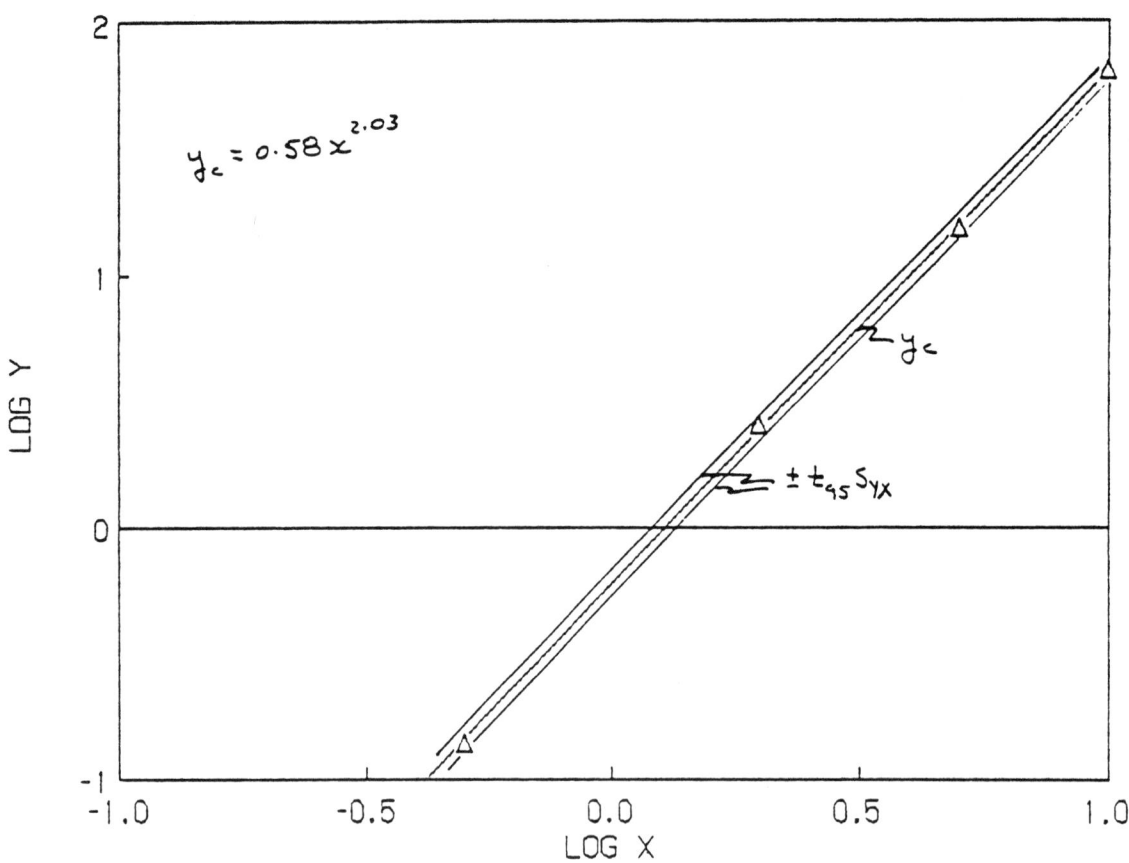

PROBLEM 4.31

KNOWN: Fan test data for Q and h

FIND: h = f(Q)

SOLUTION

Linear regression is used to find $h = f(Q)$ as an mth order polynomial of the form:

$$h = a_0 + a_1 Q + a_2 Q^2 + ... + a_m Q^m$$

To keep the regression coefficients reasonable, the polynomial will be based on Q in thousands of cms, i.e. kcms. Inspection of the data clearly shows the relation is not first order so it is not attempted.

m	a_0	a_1	a_2	a_3	r	S_{yx}	$t_{95}S_{yx}$
2	4.9886	0.2906	-0.0211	-------	0.9987	0.1246	0.396
3	5.2991	0.1468	-0.0064	-0.0004	0.9999	0.0329	0.013

Inspection of S_{yx} clearly shows the third order polynomial is preferred. The precision interval of the fit is also given. The worst deviation from the curve fit of 0.2% occurs at Q = 14 kcms. Note how the r value is not very sensitive here.

PROBLEM 4.32

KNOWN: Constraint: want $\sigma^2 \leq 1.25 \times 10^{-4}$
$S_x^2 = 2.1 \times 10^{-4}$ based on N = 30 for a large batch

FIND: Is rejection of the batch based on S_x^2 prudent?

SOLUTION

For $\nu = 29$ and using (4.25):

$$\chi^2_\alpha(\nu) = \nu S_x^2/\sigma^2 = (29)(2.1 \times 10^{-4})/1.25 \times 10^{-4} = 48.7$$

Inspection of Table 4.5 shows, $\chi^2_{0.015}(29) \approx 48.7$, so $\alpha \approx 0.015$.

So there is about a 1.5% chance that this batch actually meets the constraint despite the S_x^2 value of the sample. Rejection of the batch is prudent (or at least 98.5% prudent!).

PROBLEM 4.33

KNOWN: $\bar{x}$ = 5.060 mm with S_x = 0.0025 mm based on N = 30 measurements
Constraint: want x' = 5.000 mm based on N = 10,000 units.

FIND: Does the sample mean support that the constraint is met?

SOLUTION

For ν = 29, $t_{29,95}$ ≈ 2.045. The precision interval of the mean is given by $\pm tS_x/N^{0.5}$. Hence, this data set suggests

$$5.059 \le x' \le 5.061 \text{ mm} \quad (95\%)$$

No, the constraint is not being met. Stop the machine and reset the set-up.

PROBLEM 4.34

KNOWN: Constraint: CI = ± 0.010 mm
S_1 = 0.0025 mm based on N_1 = 30

FIND: N_T

SOLUTION

When the precision interval $\pm tS_x/N^{0.5}$ based on a reasonable number of measurements (such as 30 or more such that the t value does not change appreciably with increased N) is met, no further measurements are necessary. So $N_T = N_1$.

PROBLEM 4.35

FIND: $\bar{x}$, S_x. Test the hypothesis of a normal distribution.

SOLUTION

Tons	x_j	n_j	n'_j	$(n_j - n'_j)^2/n'_j$
5-12.9	8.95	13	10.6	0.543
13-20.9	16.95	29	26.7	0.198
21-28.9	24.95	26	28.2	0.172
29-36.9	32.95	12	9.6	0.600

$$\bar{x} = \frac{\Sigma x_i}{N} = \frac{\Sigma n_j x_j}{N} = 1652/80 = 20.65 \text{ tons}$$

$$S_x = \left[\frac{\Sigma(x_i - \bar{x})^2}{N-1}\right]^{1/2} = \left[\Sigma(n_j x_j^2 - N\bar{x}^2)/N-1\right]^{1/2}$$

$$= [(38586 - 34114)/79]^{0.5} = 7.52 \text{ tons}$$

Do the data support the hypothesis of a normal distribution? The expected occurrences n'_j are listed above. For example, for the first interval:

$$z_a = (5 - 20.65)/7.52 = 2.08 \qquad z_b = (12.9 - 20.65)/7.52 = 1.03$$

$$P(5 \le x_i \le 12.9) = P(2.08) - P(1.03) = 0.4812 - 0.3485 = 0.1327$$

For $N = 80$: $n'_1 = (80)(.1327) = 10.6$.

For all the intervals:

$$\chi^2_\alpha = \sum_j (n_j - n'_j)^2/n'_j = 1.513$$

Then, for $K = 4$, $\nu = 2$ and $\chi^2_\alpha(2) = 1.51$. Interpolation of Table 4.5 (or using a math handbook) $\alpha \approx 0.60$. There is a 60% chance that the discrepancy between n_j and n'_j is due to random variation alone (or $P = 1 - \alpha = 0.40$ or 40% chance it is not). The result is equivocal, that is it is possible and is not disproven.

PROBLEM 4.36

FIND: $\bar{x}$, S_x. Test the hypothesis of a normal distribution.

SOLUTION

Strength	x_j	n_j	n'_j	$(n_j - n'_j)^2/n'_j$
421-480	450.5	4	2.9	0.42
481-540	510.5	8	9.6	0.27
541-600	570.5	12	7.8	2.26
601-660	630.5	6	4.2	0.77

$$\bar{x} = \frac{\sum x_i}{N} = \frac{\sum n_j x_j}{N} = 550.5 \text{ MPa}$$

$$S_x = \left[\frac{\sum(x_i - \bar{x})^2}{N-1}\right]^{1/2} = \left(\frac{\sum n_j x_j^2 - N\bar{x}^2}{N-1}\right)^{1/2} = 57.53 \text{ MPa}$$

Do the data support the hypothesis of a normal distribution? The expected occurrences n'_j are listed above. For example, for the first interval:

$$z_a = (421 - 550.5)/57.53 = 2.25 \qquad z_b = (480 - 550.5)/57.53 = 1.23$$

$$P(421 \le x_i \le 480) = P(2.25) - P(1.23) = 0.4878 - 0.390 = 0.098$$

For N = 30: $n'_1 = (30)(.098) = 2.9$.

For all the intervals:

$$\chi^2_\alpha = \sum_j (n_j - n'_j)^2/n'_j = 3.72$$

Then, for K = 4, $\nu = 2$ and $\chi^2_\alpha(2) = 3.72$. Interpolation of Table 4.5 (or using a math handbook) $\alpha \approx 0.85$. There is a 85% chance that the discrepancy between n_j and n'_j is due to random variation alone (or P = 1 - α = 0.15 or 15% chance it is not). The result is equivocal, that is it is possible and is not disproven.

PROBLEM 4.37

KNOWN: For $N_1 = 30$, $\bar{x} = 550.5$ MPa, $S_1 = 57.53$ MPa

FIND: N_T required to attain CI = $\pm 0.03 \bar{x}$

SOLUTION

$$CI = \pm(0.03)(550.5 \text{ MPa}) = \pm 16.52 \text{ MPa}$$

For $d = CI/2 = 16.52$ MPa,

$$N_T = [t_{N-1,95} S_1/d]^2 = [(2.042)(57.53)/16.52]^2 = 51$$

An additional 21 measurements should be taken to meet the precision constraint. The statistics should then be recomputed to verify that the constraint is met.

PROBLEM 4.38

KNOWN: For $N_1 = 6$, $\bar{x}_1 = 71{,}327$ psi and $S_1 = 8345$ psi

FIND: N_T required to achieve CI within $0.05 x_1$ (total range)

SOLUTION

For a total range of $(0.05)(71{,}327 \text{ psi}) = 3566$ psi, CI = ± 1783 psi. With $d = CI/2 = 1783$ psi:

$$N_T = [(6)(8345)/(1783)]^2 = 145$$

An additional 139 measurements are required to reach this precision. Because N_1 is quite small relative to N_T, a reevaluation of the sample statistics after some intermediate number of measurements would be prudent.

PROBLEM 4.39

KNOWN: $CI = 0.1$ g
$S_x = 2$ g

FIND: N

SOLUTION

Let $d = CI/2 = 0.05$ g. We seek the number of measurements required to keep $tS_{\bar{x}} \leq 0.05$ g at 95%.

$$N = (tS_x/d)^2$$

If we select a large number of measurements such that $t_{N,95} = 1.96$, then

$$N \approx 6150$$

For this value, the t value remains unchanged. Thus, a large number of measurements are required due to the restriction on CI.

PROBLEM 4.40

KNOWN: $N_1 = 60$
$S_{x1} = 1.52$ V
$CI = 0.28$

FIND: N_T

ASSUMPTIONS: 95% confidence required.
S_{x1} is representative of σ.

SOLUTION

With $N_1 = 60$, $\nu = 59$ and $t_{59,95} = 2.00$. Setting $d = CI/2 = 0.14$, then

$$N_T = (tS_{x1}/d)^2 = [(2)(1.52)/0.14]^2 = 472$$

As a first estimate, at least 472 total measurements will be needed to achieve the desired precision levels. Hence, another 412 measurements are needed.

PROBLEM 4.41

FIND: $P(\chi^2(10) > 19.0)$

SOLUTION

From Table 4.5: $0.05 \leq \chi^2(10) \leq 0.025$

that is, the probability lies between 2.5 and 5%.

PROBLEM 4.42

KNOWN: Expect 0.07 breaks per meter.

FIND: p(x) for a wire of length L = 5 m.

ASSUMPTION: Model using poisson distribution

SOLUTION

From the given information, the most probable number of breaks to be expected over the 5 m length is:

$$\lambda = x' = (0.07)L = (0.07)(5) = 0.35$$

$$p(x) = 0.35^x e^{-0.35}/x!$$

x	p(x)
1	0.247
2	0.043
3	0.005
4	0.0004
10	0.0000

For example, there is a 4.3% probability of finding 2 breaks over L.

PROBLEM 4.43

KNOWN: 2 out of every 100 screws are defective, i.e. for N = 2, P(x) = .02 where x is the number of defectives.

FIND: p(x)

SOLUTION

The binomial distribution will take the form:

$$P_b = \frac{N!}{(N-x)!\,x!}\, P^x (1-P)^{N-x} = \frac{100!}{(100-x)!\,x!}\, (.02)^x (1-.02)^{100-x}$$

Using $\lambda = x' = 2$, the poisson distribution will take the form:

$$P_p = 2^x e^{-2}/x!$$

x	Pb	Pp
0	0.1326	0.1353
1	0.2707	0.2707
2	0.2734	0.2707
4	0.0902	0.0902
10	0.0000	0.0000

COMMENT

Under certain circumstances the poisson distribution will approximate the binomial distribution. This is because the former is the limiting case of the latter, a case often proven in statistics texts. When $N \Rightarrow \infty$ and $P \Rightarrow 0$ such that NP is constant, the two are exact. We see that the results above come close to this condition.

PROBLEM 4.44

KNOWN: $\lambda = 4$

FIND: p(x) using a poisson distribution model

SOLUTION

Using $\lambda = x' = 4$ as the most probable expected value,

$$p(x) = 4^x e^{-4}/x!$$

x	p(x)
1	0.0733
2	0.1465
3	0.1954
4	0.1954
5	0.1563
10	0.0053

For example, there is a 15.63% chance of observing 5 particles in a defined time interval.

PROBLEM 5.1

SOLUTION

Bias error is a constant error that shifts all measured values of a variable by a fixed amount. In effect, the sample mean value will be offset from the true mean value by this fixed amount.

Precision error leads to scatter in the measured values obtained during the measurement of a variable under otherwise fixed operating conditions. Unlike bias errors, precision errors will change in magnitude between repeated measurements bringing on the noted scatter.

Both bias and precision errors are present in any measurement. The difficult task of the engineer is to assess the probable values of these errors.

PROBLEM 5.2

SOLUTION

Bias errors are usually estimated by comparison methods. These methods include: (i) calibration, (ii) concomitant methods, (iii) interlaboratory or different facility comparisons, or (iv) experience.

Precision errors are manifested by measured data scatter and their effects on the estimate of the true value of the measured variable can be estimated statistically using the methods discussed previously in Chapter 4.

PROBLEM 5.3

SOLUTION

TRUE VALUE: The actual value of the measured variable. The value sought by measurement. Most often refers to the true mean value of the variable which would result from an infinite sampling under perfect test control.

BEST ESTIMATE: The nearest approximation of the true value that can be made with the data set available. It is based on the data set and the precision and the bias errors involved in the measurement. It is usually offered by the sample mean value and qualified by its precision interval.

MEAN VALUE: Exact statement of the mean or central tendency of a measured data set. The mean value of a finite data set is given by its sample mean value.

UNCERTAINTY: Estimate of the precision and bias errors involved in estimating the value of a variable. It is the range of probable errors which affect the outcome of a measurement.

CONFIDENCE INTERVAL: The range (or interval) of values within which the true value is expected to lie with some probability. The confidence interval is in part based on the precision interval discussed in Chapter 4 but will also include all precision errors and bias errors involved in the measurement.

PROBLEM 5.4

KNOWN: Tire pressure gauge

FIND: Difference between u_d and u_N

SOLUTION

The design-stage uncertainty is estimated by

$$u_d = (u_0^2 + u_c^2)^{0.5}$$

where u_0 is based on the instrument interpolation error and u_c is based on the instrument error usually based on the manufacturer's specifications (such as shown in Table 1.1) or past experience with the device or judgement. It may also include other known sources of error at the time of analysis.

The Nth order uncertainty is estimated by

$$u_N = (u_c^2 + \Sigma_i u_i^2)^{0.5}$$

where u_i are uncertainties related to the measurement procedure controllability. For example, u_1 might be estimated by a simple experiment whereby the fixed pressure in a single tire is repeatedly measured (say 20 times) and the outcome stated as $\pm tS_p$. Assuming that the tire pressure did not change, variations in measured pressure would be due to measurement procedure control, as well as instrument repeatability (estimated by u_0). The values for u_d and u_N differ by the errors which enter during the conduct and control of the test.

COMMENT

Under test conditions characterized by perfect control (in the sense defined in Chapter 1), we would expect that $u_N \rightarrow u_d$. However, such control can not be achieved in practice, particularly in a manufacturing setting. Because of this, u_d is not an appropriate estimate to use for stating the outcome of test results. One should consider u_d as a first guess at the lower limit of achievable uncertainty. Such information is useful for measurement test planning and measurement system selection (these two steps are part of measurement design) and form the basis for a go-no go decision on test methods.

PROBLEM 5.5

KNOWN: Micrometer
Resolution: 0.001 inch (0.025 mm)

FIND: u_d at 95% confidence

ASSUMPTIONS: Instrument error is negligible compared with error due to resolution.

SOLUTION

The interpolation error due to instrument resolution of an analog instrument is approximated by half its least increment:

$$u_0 = 0.0005 \text{ inch or } 0.0125 \text{ mm}$$

It is not unusual for this type of instrument to have mostly negligible instrument errors. However, its zero set point can only be controlled to within the precision of the resolution, so we set $u_c \approx u_0$.

Then, the design-stage uncertainty becomes

$$u_d = \pm(u_0^2 + u_c^2)^{0.5} = \pm 0.0007 \text{ inch or } 0.0180 \text{ mm } (95\%)$$

PROBLEM 5.6

KNOWN: Analog Tachometer
Resolution: 5 rpm
Accuracy: within 1% reading

FIND: u_d at 10, 500, 5000 rpm

SOLUTION

The design-stage uncertainty is

$$u_d = \pm(u_0^2 + u_c^2)^{0.5}$$

where

$$u_0 = \pm 2.5 \text{ rpm}$$

$$u_c = 1\% \text{ of reading}$$

This yields

speed [rpm]	u_c [rpm]	u_0 [rpm]	u_d [rpm]
10	0.1	2.5	±2.5
500	5	2.5	±5.6
5000	50	2.5	±50

The uncertainty increases with rotational speed. At low speeds it is dominated by the ability to read the tachometer (resolution). At higher speeds the instrument errors dominate.

COMMENT

The statement "accuracy" is a manufacturer catch-all term which is usually not well defined. The "accuracy" statement is presumed to mean that the overall errors do not exceed 1%. We suspect the term to describe the combined effects of all known elemental errors. This misnomer causes confusion. Insist on a detailed description.

PROBLEM 5.7

KNOWN: Speedometer
Resolution: 5 mph (8kph)
Accuracy: within ±4% reading

FIND: u_d at 60 mph (90 kph)

SOLUTION

The design-stage uncertainty is

$$u_d = \pm(u_0^2 + u_c^2)^{0.5}$$

where

$$u_0 = \pm 2.5 \text{ mph } (4 \text{ kph})$$

$$u_c = \pm 4\% \text{ of reading} = \pm 2.4 \text{ mph at 60 mph} = \pm 3.6 \text{ kph at 90 kph}$$

This yields

$$u_d = \pm(2.5^2 + 2.4^2)^{0.5} = \pm 3.5 \text{ mph } (95\%) = \pm 5.4 \text{ kph } (95\%)$$

COMMENT

In the United States, automobile speedometers and their connected odometers have acceptability tolerance limits set by the government at $u = \pm 4\%$ of the reading. Particularly during the fuel tight times of the 1970's, some makers of small automobiles were believed to misrepresent automobile performance by holding a tighter precision on their units while purposely building a 2 to 4% bias error into their automobile odometers. Unless the consumer actually made a careful test of the speedometer and odometer performance, usually by some comparison means such as a road sign check, they never would have reason to disbelieve their car's indicated, but inaccurate, performance. Caveat emperor!

PROBLEM 5.8

KNOWN: Temperature sensor
  Error limit: ±0.5°C
 Readout Device
  Resolution: 0.1°C
  Accuracy: 0.6°C

FIND: u_d

SOLUTION

The design-stage uncertainty is for the combined system is

$$u_d = \pm[(u_d)_R^2 + (u_d)_s^2]^{0.5}$$

where $(u_d)_R$ is the design-stage uncertainty of the readout device and $(u_d)_s$ is that of the sensor. In either case, the design-stage uncertainty is found from

$$u_d = \pm(u_0^2 + u_c^2)^{0.5}$$

Sensor

$u_0 = 0$ (i.e. no output stage in the sensor; readout is separate)

$u_c = \pm 0.5°C$

Resolution has no relevant meaning for a sensor, only for an output device.

$$(u_d)_s = (0^2 + 0.5^2)^{0.5} = 0.5°C$$

Readout Device

$u_0 = 0.05°C$ and $u_c = 0.6°C$

So that

$$(u_d)_R = (0.05^2 + 0.6^2)^{0.5} = \pm 0.6 \ °C$$

Then, the design-stage uncertainty for this combined system becomes

$$u_d = \pm(0.5^2 + 0.6^2)^{0.5} = \pm 0.8 \ °C \quad (95\%)$$

COMMENT

Although the error limit on the sensor was given in this problem, such information would be available through various publications. For example, there is a readily available ASTM standard governing the error limits on thermocouples, common temperature sensors, to which a manufacturer's product must adhere.

PROBLEM 5.9

KNOWN: Four resistors are available: two rated at $R = 500\pm 50\,\Omega$ and two rated at $R = 2000\pm 5\%\,\Omega$.

FIND: Best design combination to form $R_T = 1000\,\Omega$

SOLUTION

We can combine the resistors in series or in parallel. Consider as Case 1, a series arrangement and as Case 2, a parallel arrangement.

Case 1

$$R_T = R_1 + R_2$$

If we use the two $500\,\Omega$ resistors, then

$$(u_d)_{R1} = \pm 50\,\Omega$$

$$(u_d)_{R2} = \pm 50\,\Omega$$

But R_1 and R_2 are related functionally. The propagation of uncertainty through to R_T is estimated by

$$(u_d)_{RT} = \pm\left[\left(\frac{\partial R_T}{\partial R_1}(u_d)_{R_1}\right)^2 + \left(\frac{\partial R_T}{\partial R_2}(u_d)_{R_2}\right)^2\right]^{1/2}$$

$$= \pm[\{(u_d)_{R1}\}^2 + \{(u_d)_{R2}\}^2]^{0.5}$$

$$= \pm 71\,\Omega \quad (95\%)$$

Case 2

$$R_T = R_1 R_2/(R_1 + R_2)$$

If we use the two $2000\,\Omega$ resistors, then

$$(u_d)_{R1} = \pm 100\,\Omega$$

$$(u_d)_{R2} = \pm 100\,\Omega$$

Again R_1 and R_2 are related functionally. The propagation of uncertainty through to R_T is estimated by

$$(u_d)_{RT} = \pm\left[\left(\frac{\partial R_T}{\partial R_1}(u_d)_{R_1}\right)^2 + \left(\frac{\partial R_T}{\partial R_2}(u_d)_{R_2}\right)^2\right]^{1/2}$$

$$= \pm\left[\left(\left(\frac{R_2}{R_1+R_2} - (R_1+R_2)^{-2}R_1 R_2\right)(u_d)_{R_1}\right)^2 + \left(\left(\frac{R_1}{R_1+R_2} - (R_1+R_2)^{-2}R_1 R_2\right)(u_d)_{R_2}\right)^2\right]^{1/2}$$

$$= \pm 35\ \Omega$$

Case 2 provides the smaller uncertainty at the design-stage. We should proceed using this design.

COMMENT

Although each individual resistor in Case 2 actually has a larger absolute uncertainty than those in Case 1, we find that the weighted combination of the two resistors in Case 2 yields a lower uncertainty. This would not be immediately obvious. Our design results from a close analysis of the sensitivity of the resultant to each contributing uncertainty. The combination in Case 2 is just less sensitive to the individual uncertainties. You might try this problem using sequential perturbation methods.

PROBLEM 5.10

KNOWN: $p = 20$ kPa @ FSO (see Note below)
$(u_o)_3 = 0.01$ kPa
$(u_o)_4 = 0.001$ kPa

FIND: Select 3 1/2 or 4 1/2 digit display.

SOLUTION

A design-stage analysis is appropriate as the selection pertains to identical instruments having only different output resolutions. The instrument uncertainty is given by:

$e_L = 0.0015 \times 20$ kPa $= 0.03$ kPa
$e_H = .002 \times 20$ kPa $= 0.04$ kPa
$e_R = 0.0025 \times 20$ kPa $= 0.05$ kPa

So that: $u_c = \pm(.03^2 + .04^2 + .05^2)^{0.05} = \pm 0.071$ kPa

At FSO, the readout will display 19.99 or 19.999:

3 1/2 digit: $u_d = \pm(.07^2 + .01^2)^{.5} = 0.07$ kPa

4 1/2 digit: $u_d = \pm(.07^2 + .001^2)^{.5} = 0.071$ kPa

The uncertainty is virtually identical regardless of the resolution in the readout. Meter resolution does not affect appreciably the uncertainty.

NOTE: A typographical error in the first printing put $p = 200$ kPa. Then, if the readout is interpreted as 19.999 (10^4)Pa at FSO, the procedure and the results remain identical to the above.

PROBLEM 5.11

KNOWN: $G = f(L, T, R, \theta)$

$u_L/L = u_T/T = u_R/R = u_\theta/\theta = 0.01$

FIND: $(u_d)_G$

SOLUTION

The shear modulus is found by

$$G = 2LT/\pi R^4 \theta$$

using the best estimates of L, T, R, and θ. Its uncertainty is evaluated by

$$u_G/G = \pm [(u_L/L)^2 + (u_T/T)^2 + 16(u_R/R)^2 + (u_\theta/\theta)^2]^{.5}$$
$$= \pm [.01^2 + .01^2 + 16(.01)^2 + .01^2]^{.5} = \pm .04 \text{ or } \pm 4\%$$

Note that even if $u_L = u_T = u_\theta = 0$, u_G/G is dominated by the uncertainty in R.

PROBLEM 5.12

KNOWN: $\eta = f(T_c, T_h)$
$u_\eta/\eta \leq 0.01$
$T_h = 40°C = 313\ K$
$T_c = 20°C = 293\ K$

FIND: u_h, u_c required

SOLUTION

We will assume that the uncertainties in measuring either temperature is the same. Note that temperatures must be in absolute and that a 1 °C change equals a 1 K change.

$$u_\eta = \pm [(u_T/T_h)^2 + (u_T T_c/T_h^2)^2]^{.5}$$

Then, for a $u_\eta \leq 0.01$ requires: $u_T \leq 0.1$ K or 0.1 °C

In a laboratory environment, such an uncertainty would be attainable with considerable care and calibration. In most engineering applications, this would be very difficult to attain.

PROBLEM 5.13

KNOWN: Heat transfer from a rod is to be determined.
 $Nu = hD/k$ is the nondimensional heat transfer.
 $u_h = 150 \pm 7\%$ W/m²-K (95%)
 $u_D = 20 \pm 0.5$ mm (95% assumed)
 $u_k = 0.6 \pm 2\%$ W/m-K (95% assumed)

FIND: u_{Nu}

SOLUTION

$$Nu = f(h, D, k)$$

so that

$$u_{Nu} = \pm \left[\left(\frac{\partial Nu}{\partial h} u_h\right)^2 + \left(\frac{\partial Nu}{\partial D} u_D\right)^2 + \left(\frac{\partial Nu}{\partial k} u_k\right)^2 \right]^{1/2}$$

$$= \pm \left[\left(\frac{D}{k} u_h\right)^2 + \left(\frac{h}{k} u_D\right)^2 + \left(-\frac{hD}{k^2} u_k\right)^2 \right]^{1/2}$$

$$= \pm \left[\left(\frac{0.02}{0.6} 10.5\right)^2 + \left(\frac{150}{0.6} 0.005\right)^2 + \left(\frac{(150)(0.02)}{0.6^2} 0.012\right)^2 \right]^{1/2}$$

where $= \pm 0.4$

$u_h = \pm(0.07)(150) = \pm 10.5$ W/m²-K
$u_D = \pm 0.0005$ m
$u_k = \pm(0.02)(0.6) = \pm 0.012$ W/m-K

Then,
 $Nu = hD/k \pm u_{Nu} = 5 \pm 0.4$ (95%)

Note here the nominal values of h, D, and k are used.

PROBLEM 5.14

KNOWN: R = 30 Ω
P = 500 W
Ohmmeter
 Resolution: 1 Ω
 Accuracy: within 5% of reading
Ammeter
 Resolution: 100 mA
 Accuracy: within 0.1% of reading

FIND: $(u_d)_E$

SOLUTION

From Ohm's Law: E = IR or in terms of power, $P = I^2 R$. For the nominal values of power and resistance given expect a current, $I = (P/R)^{0.5} = 4.08$ A. Hence,

Ohmeter
$$(u_d)_R = \pm(u_0^2 + u_c^2)^{0.5}$$
$$= \pm((0.005 \times 30\Omega)^2 + (0.5\Omega)^2)^{0.5}$$
$$= \pm 0.52 \, \Omega \quad (95\%)$$

Ammeter
$$(u_d)_I = \pm(u_0^2 + u_c^2)^{0.5}$$
$$= \pm((50 \times 10^{-3} \, A)^2 + (0.001 \times 4.08 \, A)^2)^{0.5}$$
$$= \pm 50.2 \times 10^{-3} \, A \quad (95\%)$$

Then, since voltage E = f(I,R):

$$(u_d)_E = \pm\left[\left(\frac{\partial E}{\partial I}(u_d)_I\right)^2 + \left(\frac{\partial E}{\partial R}(u_d)_R\right)^2\right]^{1/2}$$
$$= \pm\left[(R(u_d)_I)^2 + (I(u_d)_R)^2\right]^{1/2}$$
$$= \pm((30\Omega \times .0502 \, A)^2 + (4.1 \, A \times 0.52 \, \Omega)^2)^{0.5}$$
$$= \pm 2.61 \, V \quad (95\%)$$

COMMENT

Compare the groups in the $(u_d)_E$ term:

Resistance: $\frac{\partial E}{\partial R}(u_d)_R = 2.05 \, V$

Current: $\frac{\partial E}{\partial I}(u_d)_I = 1.51 \, V$

Hence, the resistance measuring device contributes most to the uncertainty in voltage measurement at the design stage. Efforts to reduce the uncertainty in the resistance measurement would be a good starting place to reduce $(u_d)_E$.

Note that the units in each of the working equations are consistent. The equations would not be logical otherwise. Use of inconsistent units is a common source for mistakes by students.

PROBLEM 5.15

SOLUTION

Design-Stage Analysis: generally such an analysis is performed at a time when only information about uncertainies in measuring equipment and appropriate engineering constants required for analysis are known or estimated. The analysis assumes perfect control of the measurement process and its procedure.

Advanced-Stage Analysis: Such an analysis allows estimates of uncertainty, such as procedural control, setability of operating conditions, and repeatability of the measured variable, to be included.

PROBLEM 5.16

SOLUTION

Replication provides a measure of the control of the operating conditions and test procedure. It does this by permitting the test engineer to quantify the differences in test results obtained from duplicate tests conducted under nominally identical conditions. Repetition provides a measure of repeatability during the same test, so that subtle differences between test conduct (in a replication) are not included.

In an advanced-stage analysis, replication effects are included as a higher-order uncertainty based on estimates found by a trial of tests designed to measure such controllability. In a multiple-measurement analysis, replication effects are usually entered as a precision error evaluated from pooled statistics analysis.

PROBLEM 5.17

KNOWN: Displacement Transducer Instrument specifications
      Linearity: $e_1 = \pm 0.25\%$ reading
      Drift: $e_2 = \pm 0.05\%/^\circ C$ reading
      Repeatability: $e_3 = \pm 0.25\%$ reading
   Output Device specifications
      Resolution: 10 μV
      Accuracy: within $\pm 0.1\%$ reading
   Expect a $10^\circ C$ variation during measurements.
   Expect a nominal displacement of x = 2 cm.

FIND: $(u_d)_x$

SOLUTION

Displacement Transducer

    From the instrument specifications, we can assume that the static sensitivity of the displacement transducer is K = 5 V/5 cm = 1 V/cm. Then combining the elemental errors for this transducer and expecting a displacement of x = 2 cm and a temperature variation of up to $10^\circ C$:

$$(u_d)_{DT} = \pm (u_o^2 + u_c^2)_{DT}^{0.5}$$

where
    $(u_o)_{DT} = 0$ (not relevant)

$$(u_c)_{DT} = \pm (e_1^2 + e_2^2 + e_3^2)^{0.5}$$
$$= \pm ((0.0025 \text{x} 2)^2 + (.005 \text{x} 10^\circ C \text{x} 2)^2 + (0.0025 \text{x} 2)^2)^{0.5}$$
$$= \pm 0.01225 \text{ V}$$

    $(u_d)_{DT} = \pm 0.01225$ V

Voltmeter

$$(u_d)_E = \pm (u_o^2 + u_c^2)_E^{0.5}$$

where
    $(u_o)_{DT} = \pm 5 \times 10^{-6}$ V

$(u_c)_{DT} = \pm(2V \times .001) = \pm 0.002$ V

$(u_d)_E = \pm 0.002$ V

Then, the design-stage uncertainty in measurement system becomes:

$$(u_d)_x = \pm[(u_d)_{DT}^2 + (u_d)_E^2]^{0.5}$$
$$= \pm 0.0124 \text{ V} = \pm 0.0124 \text{ cm} \quad (95\%)$$

COMMENT

Comparison of each term in $(u_d)_x$ shows that the transducer contributes most to the uncertainty in measured displacement at the design stage. Keep in mind that at this level of uncertainty only the intrinsic instrument errors and no procedural errors are considered.

PROBLEM 5.18

KNOWN: Measurement system of Problem 5.17
  $N = 20$
  $\bar{x} = 17.2$ mm   (Note: first printing showed 172 mm)
  $S_x = 1.7$ mm   (Note: first printing showed 17 mm)

FIND: x'

ASSUMPTIONS: Measurement is sufficiently controlled such that all errors have been randomized in the measured data. That is, all errors are considered.

SOLUTION

The measurements have provided additional information about the uncertainty involved in the using this measurement system and involved in measuring this particular variable. We will approach the problem as a multiple measurement uncertainty analysis problem. Using the procedure of problem 5.17 with $\bar{x} = 1.72$ cm   (i.e. $\bar{x} = 17.2$ mm):

$$(u_d)_{DT} = \pm 0.0105 \text{ cm } (95\%)$$

$$(u_d)_E = \pm 0.0017 \text{ cm } (95\%)$$

Then, we can identify two elements of bias error at the data acquisition source: bias due to the transducer, B_{22}, and bias due to the output device, B_{24}.

$$B_{22} = (u_d)_{DT} = \pm 0.0105 \text{ cm}$$

$$B_{24} = (u_d)_E = \pm 0.0017 \text{ cm}$$

We will set the precision in both of these elements to zero.

$$P_{22} = P_{24} = 0$$

The twenty measurements are assumed to be made in a manner which best randomizes the extraneous effects on the measured variable. The measurements provide a set of finite statistics from which a precision interval in estimating the true mean value can be made. Begin by estimating the precision error due to temporal variation in the measurand:

$$P_{29} = S_{\bar{x}} = S_x/N^{0.5} = 0.038 \text{ cm}$$

with degrees of freedom, $\nu = 19$.

The measurement bias limit can be expressed as:

$$B = (B_{24}^2 + B_{22}^2)^{0.5} = 0.0108 \text{ cm}$$

and the measurement precision index is given by

$$P = 0.038 \text{ cm}$$

The uncertainty interval is found from

$$u_{\bar{x}} = \pm[B^2 + (t_{\nu,95}P)^2]^{0.5}$$
$$= \pm 0.08 \text{ cm} \quad (95\%)$$

with $t_{19,95} = 2.093$. The best estimate for mean mass displacement is

$$\bar{x}' = 17.2 \pm 0.8 \text{ mm} \quad (95\%)$$

based on the information provided.

COMMENT

Note that we could use this information as the basis for a single sample uncertainty estimate by using the test performance obtained at 17.2 mm as representative for the measurement system and procedure used and variable measured.

PROBLEM 5.19

KNOWN: Nominal pressure value to be measured is 100 psi at 70°F
Pressure transducer
   Accuracy: within 0.5% reading
Output device
   Resolution: 0.1 psi
   Repeatability: $e_1 = 0.1$ psi
   Linearity: $e_2 = 0.1\%$ reading
   Drift: $e_3 = 0.1$ psi/6 months provided $32 < T < 90°F$

ASSUMPTIONS: $K = 1$ V/psi for the transducer

FIND: $(u_d)_p$

SOLUTION

For the transducer:

$$(u_d)_t = \pm(u_c^2 + u_0^2)^{0.5} = \pm u_c = \pm 0.005 \times 100 \text{ psi} = \pm 0.5 \text{ psi}$$

For the output device

$$(u_d)_{od} = \pm(u_c^2 + u_0^2)^{0.5}$$
$$= \pm 0.18 \text{ psi}$$

where

$$u_0 = \pm 0.05 \text{ psi}$$

$$u_c = \pm(e_1^2 + e_2^2 + e_3^2)^{0.5}$$
$$= \pm(0.1^2 + .1^2 + 0.1^2)^{0.5}$$
$$= \pm 0.17 \text{ psi}$$

Note: we have assumed a value for drift equivalent to that expected over the first 6 months of operation following calibration. This value would be adjusted accordingly in an actual situation.
Then,

$$(u_d)_p = \pm(u_t^2 + u_{od}^2)^{0.5}$$
$$= \pm 0.53 \text{ psi} \quad (95\%)$$ Thus, we can expect an uncertainty of 0.5% of the reading at 100 psi due to the measurement system alone.

PROBLEM 5.20

KNOWN: $\bar{p} = 8610$ lb/ft^2 $\bar{D} = 6.1$ in. $\bar{t} = 0.22$ in.
$S_p = 273.1$ lb/ft^2 $S_D = 0.18$ in. $S_t = 0.04$ in.
$N = 10$ $N = 10$ $N = 10$

FIND: σ'

ASSUMPTIONS: Bias errors are negligible.

SOLUTION

Assuming negligible bias error, the uncertainty problem reduces to one of finding the precision errors in each of the measured variables and estimating the propagation to the resultant stress.

$P_p = S_p/N^{0.5} = 86.4$ lb/ft^2; $P_D = S_D/N^{0.5} = 0.06$ in;
$P_t = S_t/N^{0.5} = 0.01$ in.   (i.e. $P_p = S_{\bar{p}}$, $P_D = S_{\bar{D}}$, $P_t = S_{\bar{t}}$)

Since $\sigma_t = f(p, D, t)$, (watch units!)

$$P_\sigma = \pm \left[\left(\frac{\partial \sigma}{\partial p} P_p\right)^2 + \left(\frac{\partial \sigma}{\partial D} P_D\right)^2 + \left(\frac{\partial \sigma}{\partial t} P_t\right)^2 \right]^{1/2}$$

$$= \pm \left[\left(\frac{D}{2t} P_p\right)^2 + \left(\frac{p}{2t} P_D\right)^2 + \left(-\frac{pD}{2t^2} P_t\right)^2 \right]^{1/2}$$

$$= \pm \left[(1197)^2 + (1127)^2 + (6511)^2 \right]^{1/2} = 6715 \text{ lb/ft}^2$$

with ν_σ obtained from equation 5.32

$$\nu_\sigma = \frac{\left[\left(\frac{D}{2t}P_p\right)^2 + \left(\frac{p}{2t}P_D\right)^2 + \left(\frac{pD}{2t}P_t\right)^2\right]^2}{\underbrace{\frac{\left(\frac{D}{2t}P_p\right)^4}{\nu_p}}_{} + \underbrace{\frac{\left(\frac{p}{2t}P_D\right)^4}{\nu_D}}_{} + \underbrace{\frac{\left(\frac{pD}{2t^2}P_t\right)^4}{\nu_t}}_{}} = 10$$

Then, with $t_{10,95} = 2.228$ and $\bar{\sigma} = \bar{p}\bar{D}/2\bar{t}$:

$\sigma' = 119{,}366 \pm 14{,}960$ lb/ft^2 (95%)

$ = 829 \pm 104$ lb/in^2 (95%)

$ = 5.716 \pm 0.717$ MPa (95%)

PROBLEM 5.21

KNOWN: Calibration source elemental errors, K = 3

FIND: Calibration source precision index

SOLUTION

With K = 3, the source precision index is written,

$$P_1 = [P_{11}^2 + P_{12}^2 + P_{13}^2]^{0.5}$$
$$= [0.9^2 + 1.1^2 + 0.09^2]^{0.5}$$
$$= 1.424 \text{ N/m}^2$$

with degrees of freedom,

$$\nu_1 = \frac{\left(\sum_{j=1}^{3} P_{1j}^2\right)^2}{\sum_{j=1}^{3} \frac{P_{1j}^4}{\nu_{1j}}} = \frac{(0.9^2 + 1.1^2 + 0.09^2)^2}{\frac{0.9^4}{20} + \frac{1.1^4}{9} + \frac{0.09^4}{14}}$$

$$= 23$$

PROBLEM 5.22

KNOWN: Bias and precision source errors in a measurement of force.
$\overline{F} = 200$ N

FIND: F'

SOLUTION

The source errors can be combined to find the measurement bias limit and precision index. The measurement bias limit is given by

$$B = [B_1^2 + B_2^2 + B_3^2]^{0.5} = [2^2 + 4.5^2 + 3.6^2]^{0.5} = 6.1 \text{ N}$$

Likewise, the measurement precision index is given by

$$P = [P_1^2 + P_2^2 + P_3^2]^{0.5} = [0^2 + 6.1^2 + 4.2^2]^{0.5} = 7.41 \text{ N}$$

The degrees of freedom in the measurement precision index is given by (5.25),

$$\nu = \frac{[6.1^2 + 4.2^2]^2}{\frac{6.1^4}{16} + \frac{4.2^4}{18}}$$

$$= 30$$

Then, using $t_{30,95} = 2.042$,

$$u_F = \pm[B^2 + (tP)^2]^{0.5} = \pm[6.1^2 + (2.042 \times 7.41)^2]^{0.5} = 16.58 \text{ N}$$

and F' = 200 ± 16.6 N   (95%)

PROBLEM 5.23

KNOWN: $A = XY$

X and Y Instrument Accuracy: within 0.5% reading

FIND: Estimate the uncertainty in land area

SOLUTION

The land area is found by $A = XY$. Then, the most probable estimate of the mean area is:

$$\bar{A} = \bar{X}\bar{Y} = 556 \times 222 = 123\,432 \text{ m}^2$$

In the measurement of length X:

$$B_X = B_{22} = 0.005 \times 556 = 2.8 \text{ m}$$

$$P_X = P_{29} = S_X/N^{0.5} = 1.77 \text{ m}$$

In the measurement of length Y:

$$B_y = B_{22} = 0.005 \times 222 = 1.1 \text{ m}$$

$$P_y = P_{29} = S_X/N^{0.5} = 0.74 \text{ m}$$

The propagation of these errors to the resultant area is found by

$$B_A = \pm \left[\left(\frac{\partial A}{\partial X} B_X \right)^2 + \left(\frac{\partial A}{\partial Y} B_Y \right)^2 \right]^{1/2}$$
$$= \pm \left[(\bar{Y} B_X)^2 + (\bar{X} B_Y)^2 \right]^{1/2} = 873 \text{ m}^2$$

$$P_A = \pm \left[\left(\frac{\partial A}{\partial X} P_X \right)^2 + \left(\frac{\partial A}{\partial Y} P_Y \right)^2 \right]^{1/2}$$
$$= \pm \left[(\bar{Y} P_X)^2 + (\bar{X} P_Y)^2 \right]^{1/2} = 569 \text{ m}^2$$

From equation 5.32, the degrees of freedom in P_A is

$$\nu_A = \frac{\left[(\bar{Y} P_X)^2 + (\bar{X} P_Y)^2 \right]^2}{\frac{(\bar{Y} P_X)^4}{\nu_X} + \frac{(\bar{X} P_Y)^4}{\nu_Y}} = 14$$

Then, using $t_{14,95} = 2.145$,

$$u_A = \pm[B^2 + (tP)^2]^{0.5} = 1500 m^2 \text{ and}$$
$$A' = 123432 \pm 1500 m^2 \ (95\%)$$

COMMENT

This uncertainty is about 1.3 % of the measured area. The contribution from the instrument itself is barely 0.7% of area. If we assume (and barring seismic activity somewhat safely) that the measurand does not change during measurement, the rest is due to lack of control in procedure.

PROBLEM 5.24

KNOWN: $\bar{\sigma} = 1061$ psi
$S_\sigma = 22$ psi
$N = 23$

FIND: P_{29}

ASSUMPTIONS: Scatter is due to temporal variations (P_{29}).

SOLUTION

The precision in estimating the mean value of stress contains the precision error,

$$P_{29} = S_\sigma/N^{0.5} = S_{\bar{\sigma}}$$
$$= 4.6 \text{ psi}$$

with 22 degrees of freedom.

PROBLEM 5.25

KNOWN: $N = 6$ with $S_x = 1.23$ MPa
$B_x = 1.48$ MPa

FIND: u_x

SOLUTION

The uncertainty in the mean value of strength is given by

$$u_x = \pm(B_x^2 + (tS_x)^2)^{.5}$$

For $t_{5,95} = 2.571$ and $S_{\bar{x}} = 1.23 \text{ MPa}/(6)^{.5} = 0.50$ MPa:

$$u_x = \pm[1.48^2 + (2.571 \times .5)^2]^{.5} = \pm 1.96 \text{ MPa} \quad (95\%)$$

Both bias and precision errors contribute about the same.

PROBLEM 5.26

KNOWN: Standard: $B_{14} = \pm 0.5$ psi $P_{14} = 0$
Voltmeter: $B_{24} = \pm 10\ \mu V$ $P_{24} = 0$
$B_{26} = \pm .5$ psi   (see Note)
$P_{31} = S_{yx} = 0.746$ based on $\nu = 4$

FIND: u_p

SOLUTION

From the calibration data, a least squares fit yields:

$$p\ [\text{psi}] = 0.54 + 24.03E\ [\text{mV}] \pm (2.776)(.746)$$

so that, $P_{31} = S_{yx} = 0.746$ based on $\nu = 4$ and $K = dp/dE = 24.03$ psi/mV.

$$B = (.5^2 + (0.01\ \text{mV} \times 24.03\ \text{psi/mV})^2 + .5^2)^{.5} = \pm .707\ \text{psi}$$

and

$$u_p = \pm(.707^2 + [2.776 \times .746]^2)^{.5} = \pm 2.19\ \text{psi}\quad (95\%)$$

NOTE: In the first printing, $B_{26} = \pm 5$ psi. The solution procedure remains the same. B_{26} dominates to yield, $u_p = \pm 5$ psi.

PROBLEM 5.27

KNOWN: Density of metal composite is determined by mass estimation.
Sample ingot is cylindrical in shape.
Nominal values for typical ingot:

$m \approx 4.5 \text{ lb}_m$ $(u_0)_m = 0.1 \text{ lb}_m = u_m$
$L \approx 6 \text{ in.}$ $(u_0)_L = 0.05 \text{ in.} = u_L$
$D \approx 4 \text{ in.}$ $(u_0)_D = 0.0005 \text{ in.} = u_D$

FIND: $(u_0)_\rho$

SOLUTION

$$V = \pi D^2 L / 4$$

$$\rho = m/V = 4m/\pi D^2 L$$

$$(u_0)_\rho = \pm \left[\left(\frac{\partial \rho}{\partial m} u_m\right)^2 + \left(\frac{\partial \rho}{\partial D} u_D\right)^2 + \left(\frac{\partial \rho}{\partial L} u_L\right)^2 \right]^{1/2}$$

$$= \pm \left[\left(\frac{4}{\pi D^2 L} u_m\right)^2 + \left(\frac{-8m}{\pi D^3 L} u_D\right)^2 + \left(\frac{4m}{\pi D^2 L^2} u_L\right)^2 \right]^{1/2}$$

$$= \pm 0.0013 \text{ lb}_m/\text{in}^3 = \pm 2.24 \text{ lb}_m/\text{ft}^3$$

The mass measurement contributes most to the uncertainty at the zero order level and would be improved first to reduce the uncertainty in density.

PROBLEM 5.28

KNOWN: M = 3 replications with N = 10 repetitions each.
Sample mean and sample standard deviation values.

FIND: $<\bar{d}> \pm u_{\bar{d}}$

ASSUMPTIONS: For the micrometer, $u_c \approx u_0$ from Problem 5.27

SOLUTION

We will assume that errors enter only at the data acquisition stage. Elemental errors from data acquisition sources will consist at least as a bias due to instrument error (B_{22}), a precision error due to the variation in readings measured at each cross-section (P_{25}), and spatial errors along the ingot length (P_{28}).

The instrument error is estimated by the design stage uncertainty of the measurement. It is reasonable to assume that the instrument is correct to within its interpolation error, see Example 5.14, so with $u_0 = \pm 0.0005$ in, we set $u_c = u_0$. Thus, design-stage uncertainty becomes ± 0.0007 in. We set $B_{22} = 0.0007$ in.

The pooled mean diameter is found by

$$<\bar{d}> = \frac{1}{3} \sum_{m=1}^{3} \bar{d}_m = 3.9924 \text{ in.}$$

which provides the most probable estimate in the true mean diameter. The variation in readings taken at each cross-section location is estimated by the pooled standard deviation relative to the pooled mean,

$$<S_{\bar{d}}> = <S_d>/(MN)^{0.5} = \left(\sum_{m=1}^{3} S_{d_m}^2 / 3\right) / \sqrt{30} = 0.00055$$

with $\nu_{25} = \Sigma_j (N_j - 1) = 27$. This yields a measure of P_{25}.

The spatial variation in mean values is estimated by the standard deviation

$$S_d = \left[\frac{\sum_{m=1}^{3} (\bar{d}_m - <\bar{d}>)^2}{m-1}\right]^{1/2} = 0.0034$$

with $\nu = M - 1 = 2$. A precision estimate in the mean value then is made as

$$P_{28} = S_d/M^{0.5} = 0.002 \text{ in.}$$

Then,

$$B = B_2 = B_{22} = 0.0007 \text{ in.}$$

$$P = P_2 = [P_{25}^2 + P_{28}^2]^{0.5} = 0.002 \text{ in.}$$

with

$$\nu = \frac{[0.002^2 + 0.00055^2]^2}{\frac{0.002^4}{2} + \frac{.00055^4}{27}} = 2.5 \approx 2$$

such that $t_{2,95} = 4.303$ (note: ν is rounded down here).

$$u = \pm(B^2 + (tP)^2)^{0.5} = \pm 0.0086 \text{ in.}$$

$$B = 0.0007 \text{ in.} \quad P = 0.002 \text{ in.} \quad \nu = 2$$

$$d' = 3.9924 \pm 0.0086 \text{ in.} \quad (95\%)$$

COMMENT

We can see that the variation in the ingot diameter between the three cross-sections contributes the most to the uncertainty in stating a true mean ingot diameter. With such a large variation in d, M needs to be larger to improve the confidence in the estimated mean. Increasing N would not improve the confidence in the mean value estimate significantly in this problem.

This problem provides a nice example of how the statement of a value of a variable such as the diameter of an ingot or of a shaft is actually a statistical statement. We often think of such non-temporal variables as being fixed and absolute. Because they are statistical statements, there is an associated uncertainty in their values.

PROBLEM 5.29

KNOWN: Measurements of mass and length and results of Problem 5.28

FIND: ρ'

SOLUTION

We will assume that errors enter only at the data acquisition stage. Elemental errors from data acquisition sources will consist at least as a bias due to instrument error (B_{22}), a precision error due to the variation in readings (P_{25}). We will also use the bias and precision estimates from the diameter measurement information in Problem 5.28. These give:

$$(B)_d = 0.0007 \text{ in.}, \quad (P)_d = 0.0016 \text{ in.}, \quad \nu = 2$$

The instrument error is estimated by the design stage uncertainty of each measured variable. We will assume design-stage uncertainty values using $u_c \approx u_0$ and assign these values as bias errors, B_{22}.

$$(B_{22})_m = 0.14 \text{ lb}_m \quad (B_{22})_L = 0.071 \text{ in.}$$

The variation in measured readings affects the precision in the estimated mean value. The precision estimates in the mean values are assumed as

$$(P_{25})_m = S_m/N^{0.5} = 0.022 \text{ lb}_m \quad \nu_m = 20$$
$$(P_{25})_L = S_L/N^{0.5} = 0.0302 \text{ in.} \quad \nu_L = 10$$

Then the measurement bias and precision estimates are:

$(P)_d = 0.0016$ in. $\nu = 2$; $(P)_m = (P_{25})_m = 0.022 \text{ lb}_m$ $\nu = 20$;
$(P)_L = (P_{25})_L = 0.0302$ in. $\nu = 10$
$(B)_d = 0.0007$ in. ; $(B)_m = (B_{22})_m = 0.14 \text{ lb}_m$; $(B)_L = (B_{22})_L = 0.071$ in.

These errors are propagated to the resultant density through the relation $\rho = 4m/\pi d^2 L$.

$$(B)_\rho = \pm\left[\left(\frac{\partial \rho}{\partial D} B_D\right)^2 + \left(\frac{\partial \rho}{\partial m} B_m\right)^2 + \left(\frac{\partial \rho}{\partial L} B_L\right)^2\right]^{1/2}$$
$$= \pm[(2\times10^{-5})^2 + (1.9\times10^{-3})^2 + (7\times10^{-4})^2]^{0.5} = 0.002 \text{ lb}_m/\text{in}^3$$

$$(P)_\rho = \pm\left[\left(\frac{\partial \rho}{\partial D} P_D\right)^2 + \left(\frac{\partial \rho}{\partial m} P_m\right)^2 + \left(\frac{\partial \rho}{\partial L} P_L\right)^2\right]^{1/2}$$
$$= \pm[(5\times10^{-4})^2 + (3\times10^{-4})^2 + (3\times10^{-4})^2]^{0.5} = 4.2 \times 10^{-4} \text{ lb}_m/\text{in}^3$$

where each term is evaluated at the mean values for m, L and d.

$$(\nu)_\rho = \frac{\left((5\times 10^{-4})^2 + (3\times 10^{-4})^2 + (3\times 10^{-4})^2\right)^2}{\frac{(5\times 10^{-4})^4}{2} + \frac{(3\times 10^{-4})^4}{20} + \frac{(3\times 10^{-4})^4}{10}} = 22$$

so that: $u_\rho = \pm[(B)^2_\rho + (tP)_\rho^2]^{0.5} = \pm[0.002^2 + (2.07 \times 0.0004)^2]^{0.5} = 0.0022$

$$\rho' = 4\bar{m}/\pi \bar{d}^2 \bar{L} \pm u_\rho = 0.0602 \pm 0.0022 \text{ lb}_m/\text{in}^3 \text{ (95\%)}$$

or $u_\rho/\rho \approx 3\%$. The instrument bias error in mass contributes the most to the uncertainty.

PROBLEM 5.30

KNOWN: Calibration against a standard.
Calibration data are provided.

FIND: a.) Calibration curve fit. b.) Uncertainty in any estimated T.

SOLUTION

a.) To compute a calibration curve fit, the data are fit to a polynomial using a least squares analysis (such as in Appendix A). An acceptable first order curve fit is obtained: $E = -0.022 + 0.042T \pm 0.099$ mV (95%) with $t_{4,95} = 2.770$. This corresponds to the curve: $T = 0.564 + 24E \pm 0.83°C$ (95%) with $S_{yx} = 0.3°C$.

b.) From the problem statement, we can set:
$$B_{14} = 0.05°C$$
$$B_{15} = 5°C/m \times 0.010m = 0.05°C$$

Each value of independent variable is measured once and the results combined to form the curve fit in part a. Voltage measurement system bias errors are not indicated but will be present in the measured data. We will make the assumption that such errors are negligible to within the resolution of the instrument, so that we set:
$$B_{24} = 0.001 mV$$

(Note: from experience or manufacturer specifications, we might change this value up or down). From part a, we see that the temperature system has a static sensitivity of 2.38°C/mV, so that $B_{24} = 0.002°C$, a negligible amount. Precision errors in the measurement system are presumed included in the calibration curve.

Lastly, the precision error in the calibration curve is estimated as $P_{31} = S_{yx} = 0.3°C$ with $\nu = 4$. Hence,

$$B = [B_1^2 + B_2^2 + B_3^2]^{0.5} = [(0.05^2 + 0.05)^2 + (0.002)^2 + 0^2]^{0.5}$$
$$= \pm 0.071°C$$

$$P = 0.3°C \quad \nu = 4$$

$$u_T = [B^2 + (tP)^2]^{0.5} = [0.071^2 + (2.770 \times 0.3)^2]^{0.5} = 0.8°C \ (95\%)$$

Here the data scatter on the curve fit has the greatest effect on the uncertainty.

PROBLEM 5.31

KNOWN: $P = E^2/R$

$R \approx 100\,\Omega \quad P \approx 100\,W$

Ohmmeter
 Resolution: $1\,\Omega$
 Error: 1% reading

Voltmeter
 Resolution: $1\,V$
 Error: 1% reading

FIND: u_0 and u_d for power

SOLUTION

using $E = [(100\,\Omega)(100\,W)]^{1/2} = 100\,V$:

Zero-order uncertainty

$(u_0)_E = \pm 0.5\,V \qquad (u_0)_R = \pm 0.5\,\Omega$

$$(u_0)_P = \pm \left[\left(\frac{2E}{R}(u_0)_E\right)^2 + \left(\frac{-E^2}{R^2}(u_0)_R\right)^2 \right]^{1/2}$$

$= \pm 1.12\,W \quad (95\%)$

Design-stage uncertainty

$$u_d = [u_0^2 + u_c^2]^{0.5}$$

$(u_c)_E = 1\,V$
$(u_c)_R = 1\,\Omega$

$(u_d)_E = [0.5^2 + 1^2]^{0.5} = 1.12\,V$
$(u_d)_R = [0.5^2 + 1^2]^{0.5} = 1.12\,\Omega$

$$(u_d)_P = \pm \left[\left(\frac{2E}{R}(u_d)_E\right)^2 + \left(\frac{-E^2}{R}(u_d)_R\right)^2 \right]^{1/2}$$

$= \pm 2.5\,W \quad (95\%)$

COMMENT

The uncertainty is updated as each new piece of information is added into the analysis.

PROBLEM 5.32

KNOWN: P = 10, 1000, 10000 W
$P = E^2/R$ or $P = IE$
Instrument specifications in Table

FIND: Design (select) a best method using uncertainty analysis

SOLUTION

This problem is open-ended and this solution is offered only as one guide. Suppose we fix E = 100V. This determines R and I for analysis.

$(u_d)_E = [0.5^2 + (0.005 \times E)^2]^{0.5}$
$(u_d)_A = [0.25^2 + (0.01 \times A)^2]^{0.5}$
$(u_d)_R = [0.5^2 + (0.005 \times R)^2]^{0.5}$

Method 1: $P = E^2/R$

$$(u_d)_P = \pm \left[\left(\frac{\partial P}{\partial E}(u_d)_E\right)^2 + \left(\frac{\partial P}{\partial R}(u_d)_R\right)^2 \right]^{1/2}$$

or

$$(u_d)_P/P = \pm \left[\left(\frac{2}{E}(u_d)_E\right)^2 + \left((u_d)_R/R\right)^2 \right]^{1/2}$$

P [W]	E [V]	R [Ω]	u_P/P [%]
10	100	1000	1.5
100	100	10	2.9
1000	100	1	50.0

Method 2: $P = EI$

$$(u_d)_P = \pm \left[\left(\frac{\partial P}{\partial I}(u_d)_I\right)^2 + \left(\frac{\partial P}{\partial E}(u_d)_E\right)^2 \right]^{1/2}$$

or

$$(u_d)_P/P = \pm \left[\left((u_d)_I/I\right)^2 + \left((u_d)_E/E\right)^2 \right]^{1/2}$$

P [W]	E [V]	I [A]	u_P/P [%]
10	100	0.1	250
100	100	10	2.8
1000	100	100	1.3

Method 1 is better at low power levels while method 2 is better at high levels.

COMMENT

Different values of E will produce different results. A broader look of this problem would vary E and optimize to determine a basis for preferred operating conditions (in E, R, and I).

PROBLEM 5.33

KNOWN: Composite material is function of cure temperature, $\sigma = f(T)$.
Possible cure temperature range: 20^0C to 60^0C
Oven controllability to be tested at 30^0C:
  Oven divided into quadrants: $j = 1$ to J
  N measurements taken in each quadrant: $i = 1$ to N
  M replications to be made of entire test: $m = 1$ to M
  Method 1: $N = 5$, $M = 5$, $J = 4$ i.e. JxNxM = 100
  Method 2: $N = 25$, $M = 1$, $J = 4$ i.e. JxNxM = 100

FIND: If uncertainty in the oven test temperature is to be estimated, discuss information obtained by the two different methods.

ASSUMPTIONS: In order to simplify the solution we assume that sensor installation effects and measurement system operating conditions are properly controlled. We also neglect data reduction errors. In both cases the effects will be the same for either method.

SOLUTION

Our goal is to estimate the uncertainty associated with the $\sigma(T)$ test due to controllability of the independent variable, T. This test is really one of oven performance. By setting the oven to one representative condition, we can estimate typical oven performance at other temperatures through a single measurement analysis.

At any set temperature:

(i) By dividing the oven into quadrants and determining the mean quadrant temperature, we obtain information concerning the typical oven spatial variation in temperature.

(ii) By repetition, we obtain information about the typical oven temporal variation in temperature.

(iii) By replication, we obtain information about our ability to repeat the exact conditions on subsequent attempts (obviously, variations in the actual achieved oven temperature, despite the fact that the oven controls are seemingly reset exactly the same on each replication, affects strength).

Define:
  u_1: temporal variation effect on mean oven temperature
  u_2: spatial variation effect on mean oven temperature
  u_3: set controllability (repeatability) of the mean oven temperature
  u_c: instrument error associated with the measuring equipment

Mean temperature of any quadrant on any replication: $\overline{T}_{jm}$

$$\overline{T}_{jm} = \frac{1}{N} \sum_{i=1}^{N} T_{ijm}$$

Grand pooled mean temperature: $\langle \overline{\overline{T}} \rangle$

$$\langle \overline{\overline{T}} \rangle = \frac{1}{M} \sum_{m=1}^{M} \langle \overline{T}_m \rangle$$

Pooled mean oven temperature on any replication: $\langle \overline{T}_m \rangle$

$$\langle \overline{T}_m \rangle = \frac{1}{J} \sum_{j=1}^{J} \overline{T}_{jm}$$

Spatial temperature variation on any replication: S_{T_m}

$$S_{T_m} = \left[\frac{1}{J-1} \sum_{j=1}^{4} (\overline{T}_{jm} - \langle \overline{T}_m \rangle)^2 \right]^{1/2}$$

Temporal temperature variation on any replication:

$$\langle S_{T_m} \rangle = \left[\frac{1}{J(N-1)} \sum_{j=1}^{4} \sum_{i=1}^{N} (T_{ijm} - \langle \overline{T}_m \rangle)^2 \right]^{1/2}$$

Set controllability of mean temperature:

$$\langle S_{\overline{\overline{T}}} \rangle = \left[\frac{1}{M-1} \sum_{m=1}^{M} (\langle \overline{T}_m \rangle - \langle \overline{\overline{T}} \rangle)^2 \right]^{1/2}$$

$$u_N = \pm [u_c^2 + u_1^2 + u_2^2 + u_3^2]^{0.5} \quad (P\%)$$

Method 1:   Method 2:

$u_1 = t \langle S_{T_m} \rangle / (JN)^{1/2}$ $u_1 = t \langle S_{T_m} \rangle / (JN)^{1/2}$

$u_2 = t \, S_{T_m} / (J)^{1/2}$ $u_2 = t \, S_{T_m} / (J)^{1/2}$

$u_3 = t \langle S_{\overline{\overline{T}}} \rangle / (M)^{1/2}$ u_3 = information not available without replication

We see that without replication we will have no information on our ability to control the oven set temperature. This will be an important omission if we intend to use the oven for batch production.

COMMENT

There may be other options as to how to approach this problem and the above analysis presents but one logical solution.

As an alternative, we could perform a multiple measurement analysis to estimate our ability to control the oven at 30°C. Using,

$B_{22} = u_c$ $P_{25} = \langle S_{\overline{\overline{T}}} \rangle$ $P_{29} = \langle S_{T_m} \rangle$ $P_{28} = S_{T_m}$

we should obtain about the same result.

PROBLEM 5.34

KNOWN: $N_1 = 50$
$\bar{x} = 2.112$ V
$S_1 = 0.387$
Want CI = 0.10 at 95%

FIND: N_T

SOLUTION

$$CI = \pm tS/N^{0.5}$$

Let $d = CI/2$, then

$$N_T \approx (t_{N_1-1,95}S_1/d)^2$$
$$\approx (2.01 \times 0.387/0.05)^2 = 242$$

PROBLEM 5.35

KNOWN: N = 10 measurements made at M = 4 locations
Micrometer: Resolution: 0.001 in.; Accuracy: < 0.001 in.

FIND: D'

ASSUMPTIONS: We will restrict the solution to errors due to data acquisition sources.

SOLUTION

A pooled estimate of the diameter of the shaft is given by

$$\langle \bar{D} \rangle = \frac{4.494 + 4.499 + 4.511 + 4.522}{4} = 4.506$$

with standard deviation

$$\langle S_D \rangle = \left[\frac{.006^2 + .009^2 + .01^2 + .003^2}{4} \right]^{1/2} = 0.0075 \text{ in.}$$

We can recognize elemental errors due to instrument error and due to spatial variations effects on the computed mean value and procedural variation errors which bring about data scatter at each cross section. Instrument errors will be treated as a bias error,

$$B_{22} = u_c = 0.001 \text{ in.}$$

This value reflects the manufacturer accuracy statement. The procedural variations bring on a precision error in each measured mean value as estimated by

$$P_{25} = \langle S_D \rangle / (MN)^{0.5} = 0.0012 \text{ in.} \quad \Rightarrow \quad P_{25} = \langle S_{\bar{D}} \rangle$$

with degrees of freedom, $\nu_{25} = M(N_j - 1)$. The spatial error arises because the diameter is not uniform along its length such that the mean values vary. This spatial error affects the pooled mean value and is estimated by

$$S_{\bar{D}} = \left[\frac{\sum_{j=1}^{4} (\bar{D}_j - \langle \bar{D} \rangle)^2}{3} \right]^{1/2} = 0.0126 \text{ in}$$

$$P_{28} = S_{\bar{D}} / M^{0.5} = 0.006 \text{ in.} \quad \Rightarrow \quad P_{28} = S_{\bar{\bar{D}}}$$

Take care of the nomenclature for $S_{\bar{D}}$ above.

Collecting precision terms,

$$P = [P_{25}^2 + P_{28}^2]^{0.5} = 0.006 \text{ in.}$$

with degrees of freedom

$$\nu_D = \frac{[.001^2 + .006^2]^2}{\frac{.001^4}{36} + \frac{.006^4}{3}} \approx 3$$

The uncertainty estimate can be given by

$$u_D = [B^2 + (t_{3,95}P)^2]^{0.5} = [.001^2 + (3.182 \times 0.006)^2]^{0.5}$$

$$D' = 4.506 \pm 0.019 \text{ in} (95\%)$$

PROBLEM 5.36

KNOWN: Pressure is measured using a dial gauge.
$\bar{p} = 50$ psi $S_p = 2$ psi $M = 30$ $N = 1$
Dial gauge:
  Resolution: 0.1 psi
  Accuracy: within 0.5 psi

FIND: Estimate the uncertainty in vessel pressure.

SOLUTION

Single measurement analysis: From the manufacturer statement, we set

$$u_c = 0.5 \text{ psi}$$

The variations in vessel pressure upon each replication are estimated by

$$S_p = 2 \text{ psi}$$

such that,

$$u_1 = t_{29,95} S_p = (2.047)(2) = 4.094 \text{ psi} = 4.1 \text{ psi}$$

Then the uncertainty in vessel set pressure at any time can be estimated by

$$u_N = \pm[0.5^2 + 4.1^2]^{0.5} = \pm 4.1 \text{ psi} \quad (95\%)$$

PROBLEM 5.37

KNOWN: Transducer, readout specifications.
$M = 4$ replications with $N = 10$ repetitions each.

FIND: $\bar{x} \pm u_x$

SOLUTION

From the data, $\langle \bar{x} \rangle = (4.3 + 3.8 + 4.2 + 4.0)/4 = 4.08$

At the design-stage, only information known prior to the test are included:

For the transducer:
$$e_L = \pm (0.0025)(4m) = 0.01 \text{ m}$$
$$e_R = \pm (0.0025)(4m) = 0.01 \text{ m}$$
$$e_K = \pm (0.001)(5m) = 0.005 \text{ m}$$
$$e_z = \pm (0.005/^\circ C)(3^\circ C)(5m) = 0.0075 \text{ m}$$
so that,
$$(u_c)_t = \pm (.01^2 + .01^2 + .005^2 + .0075^2)^{0.5} = 0.0167 \text{ m}$$

The transducer has a sensitivity $K_t = 1V/m$ so that the voltmeter output can be stated in terms of displacement [m].

For the voltmeter:
$$(u_d)_E = \pm \{[(0.001)(4m)(1m/V)]^2 + [(5\mu V)(1m/V)]^2\}^{0.5} = \pm 0.004 \text{ m}$$

For the transducer-voltmeter system, the design-stage uncertainty is:
$$u_d = \pm (.004^2 + .0167^2)^{0.5} = 0.017 \text{ m} \quad (95\%)$$

We can use this information for the multiple-measurement analysis where we assign instrument errors:

$$B_{22} = (u_c)_t = 0.0167 \text{ m} \qquad B_{24} = (u_c)_E = 0.004 \text{ m}$$

The uncertainty in the applied input estimates the control of the input forcing function. Assuming $F = kx$, then $K = 2000 \text{ N}/4 \text{ m} = 500 \text{ N/m}$. Hence, we assign:

$$B_{25} = (dx/dF)u_F = 0.2 \text{ m}$$

The temporal or repetition error in the mean values is estimated by:

$$P_{29} = [(S_1^2 + S_2^2 + S_3^2 + S_4^2)/4]^{0.5}/(MN)^{0.5} = 0.04 \text{ m} \text{ with } \nu_{29} = 36$$

The temporal error or replication error in the overall pooled mean is:

$$P_{28} = [\sum_j (\bar{x}_j - <\bar{x}>)^2/3]^{0.5}/(M)^{0.5} = 0.11 \text{ m} \text{ with } \nu_{28} = 4$$

Using (5.25), $\nu = f(\nu_{29}, \nu_{28}) = 4$

Then, for the measurement:

$$B = \pm (.0167^2 + .004^2 + .2^2)^{0.5} = \pm 0.2 \text{ m}$$

The ability to control the applied force dominates.

$$P = \pm (.11^2 + .04^2)^{0.5} = \pm 0.12 \text{ m}$$

Then, $u_x = \pm (.2^2 + [(2.770)(.12)]^2)^{0.5} = \pm 0.39 \text{ m}$ (95%)

Hence, $x' = <\bar{x}> \pm u_x = 4.08 \pm 0.39 \text{ m}$ (95%)

PROBLEM 5.38

KNOWN: First-order system:
$$u_\Gamma/\Gamma = \pm 0.02 \ (95\%)$$
$$u_t/t = \pm 0.01 \ (95\%)$$

FIND: u_τ/τ

ASSUMPTION $\tau = f(\Gamma, t)$ with no other influences.

SOLUTION

$$\Gamma = e^{-t/\tau}$$

Rearranging, $\tau = -t/\ln \Gamma$

$$u_\tau = \pm [(u_t/\ln\Gamma)^2 + (tu_\Gamma/\Gamma\{\ln\Gamma\}^2)^2]^{0.5}$$

$$u_\tau/\tau = \pm [(u_t/t)^2 + (u_\Gamma/\Gamma\ln\Gamma)^2]^{0.5}$$

$$= \pm [.01^2 + (.02/\ln\Gamma)^2]^{0.5}$$

Γ	u_τ/τ
.1	0.013
.5	0.031
.8	0.090
.9	0.190
1.0	∞

PROBLEM 5.39

KNOWN: Transducer, readout specifications.
$M = 3$ replications with $N = 20$ repetitions each.

FIND: $\bar{T} \pm u_T$

SOLUTION

From the data, $<\bar{T}> = (181.0 + 183.1 + 182.1)/3 = 182.1\ ^\circ C$

From the problem statement, we can assign
$B_{22} = e_{22} = \pm 1\ ^\circ C$ measuring system error
$B_{26} = e_{26} = \pm 1.2\ ^\circ C$ installation error

The temporal or repetition error in the mean values is estimated by:

$$P_{29} = [(S_1^2 + S_2^2 + S_3^2)/3]^{0.5}/(MN)^{0.5} = 0.38\ ^\circ C \text{ with } \nu_{29} = 57$$

The temporal error or replication error in the overall pooled mean is:

$$P_{21} = [\Sigma\,(\bar{T}_j - <\bar{T}>)^2/2]^{0.5}/(M)^{0.5} = 0.61\ m \text{ with } \nu_{21} = 2$$

Using (5.25), $\nu = f(\nu_{29}, \nu_{21}) = 3.8 \approx 4$

Then, for the measurement:

$$B = \pm (1^2 + 1.2^2)^{0.5} = \pm 1.56\ ^\circ C$$

$$P = \pm (.38^2 + .61^2)^{0.5} = \pm 0.72\ ^\circ C$$

Then, $u_T = \pm (1.56^2 + [(2.770)(.72)]^2)^{0.5} = \pm 2.5\ ^\circ C$ (95%)

Hence, $T' = <\bar{T}> \pm u_T = 182.1 \pm 2.5\ ^\circ C$ (95%)

PROBLEM 5.40

SOLUTION

Exact answers will depend on user experience. As a class exercise, it would be interesting to discuss why different people chose their numbers. As a guide:

bathroom scale:
    spring scale: ± 5 to 10 N
    balance beam scale: ± 1 N

plastic ruler: to within its resolution

micrometer: to within its resolution

kitchen thermometer: ± 1 °C

speedometer: ± 4% reading   (in USA this is by law)

PROBLEM 5.41

KNOWN: Air @ T = 25 ± 2 °C (95%)
$u_p/p = 0.01$ (95%)

FIND: u_ρ/ρ

ASSUMPTIONS: Ideal gas $p = \rho RT$. Neglect u_R.

SOLUTION

$$u_\rho/\rho = [(u_p/p)^2 + (u_T/T)^2]^{0.5}$$

A change in 1 °C equals a change of 1 K. So

$T = (25 + 273) \pm 2$ K

$u_\rho/\rho = [(0.01)^2 + (2\text{ K}/298)^2]^{0.5} = \pm 0.012$ or 1.2% (95%)

PROBLEM 5.42

KNOWN: Air with $u_T = \pm 1\ ^\circ C$, $u_\rho/\rho = \pm 0.005$

FIND: u_p/p

ASSUMPTIONS: Ideal gas $p = \rho RT$. Neglect u_R. Air at 25 $^\circ$C.

SOLUTION

$$u_p/p = \pm [(u_\rho/\rho)^2 + (u_T/T)^2]^{0.5}$$

A change in 1 $^\circ$C equals a change of 1 K. So, T = 298 ± 1 K.

$$u_p/p = \pm [(0.005)^2 + (1/298)^2]^{0.5} = \pm 0.0037 = \pm 3.7\%\quad (95\%)$$

PROBLEM 5.43

KNOWN: $T = e^{-KEW}$ $0 \leq T \leq 1$
$E = 2 \pm 0.04$ (95%) (see note below)
$u_T/T = 0.01$ (95%)
$u_W/W = 0.01$ (95%)

FIND: Is $u_K/K \leq 0.05$? 0.10?

SOLUTION

The solids density is found by

$$K = -(\ln T)/EW$$

so that,

$$u_K/K = \pm [(u_T/T\ln T)^2 + (u_E/E)^2 + (u_W/W)^2]^{0.5}$$
$$= \pm [(0.01/\ln T)^2 + (.04/2)^2 + (.01)^2]^{0.5} \quad (95\%)$$

T	u_K/K
.5	0.027
.8	0.05
.9	0.0975
1.	∞

So u_K/K approaches 5% at $T = 0.8$ and 10% at $T = 0.9$. $u_K/K \to \infty$ as $T \to 1$.

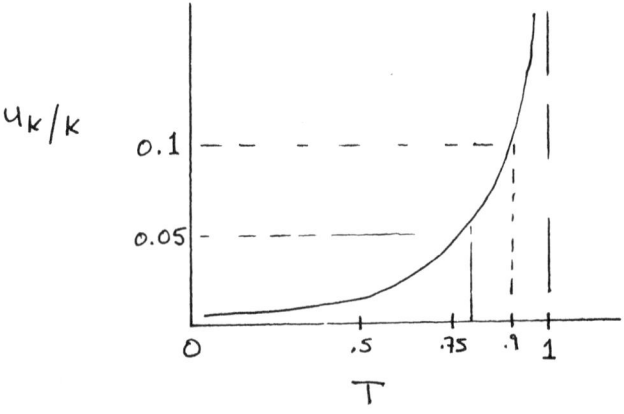

PROBLEM 5.44

KNOWN: First-order system

$u_{\Gamma}/\Gamma = \pm 0.02$ (95%)
$u_t/t = \pm 0.005$ (95%)

FIND: u_τ/τ

SOLUTION

$$\Gamma = e^{-t/\tau}$$

Rearranging, $\tau = -t/\ln \Gamma$. Then,

$$u_\tau = \pm [(u_t/\ln\Gamma)^2 + (tu_\Gamma/\Gamma\{\ln\Gamma\}^2)^2]^{0.5}$$

$$u_\tau/\tau = \pm [(u_t/t)^2 + (u_\Gamma/\Gamma\ln\Gamma)^2]^{0.5}$$

$$= \pm [.005^2 + (.02/\ln\Gamma)^2]^{0.5}$$

Γ	u_τ/τ
.1	0.010
.5	0.029
.8	0.090
.9	0.190
1.	∞

Because τ approaches steady state asymptotically as Γ goes to 1, this behavior is not unexpected.

PROBLEM 6.1

KNOWN: A current loop having

$N = 20$
$A = 1 \text{ in}^2 = 0.000645 \text{ m}^2$
$I = 0.02 \text{ A}$
$= 0.4 \text{ Wb/m}^2$

FIND: Torque on the current loop, T_μ.

ASSUMPTIONS: The magnetic field is oriented at an angle of 90° to the current flow direction.

SOLUTION:

From (6.3)
$$T_\mu = NIAB \sin \alpha$$

The maximum torque occurs when $\sin \alpha = 1$, which yields

$$T_\mu^{max} = NIAB$$
$$= 20(0.02 \text{ A})(0.000645 \text{ m}^2)(0.4 \text{ Wb}/\text{m}^2)$$

$$T_\mu^{max} = 1.03 \times 10^{-4} \text{ N-m}$$

PROBLEM 6.2

KNOWN: A voltage dividing circuit with $R_T = 5000\ \Omega$ (as shown in Figure 6.16)

FIND: R_m such that the loading error is less than 12% of the full scale value.

SOLUTION: Since the full scale output is E_i (with e as the error)

$$\frac{e}{E_i} = \frac{R_1 - R_T + (R_T - R_1)\left(R_1/R_m + 1\right)}{R_T + \left(R_T^2/R_1 - R_T\right)\left(R_1/R_m + 1\right)}$$

A plot of the error as a function of meter resistance and R_1 is shown below

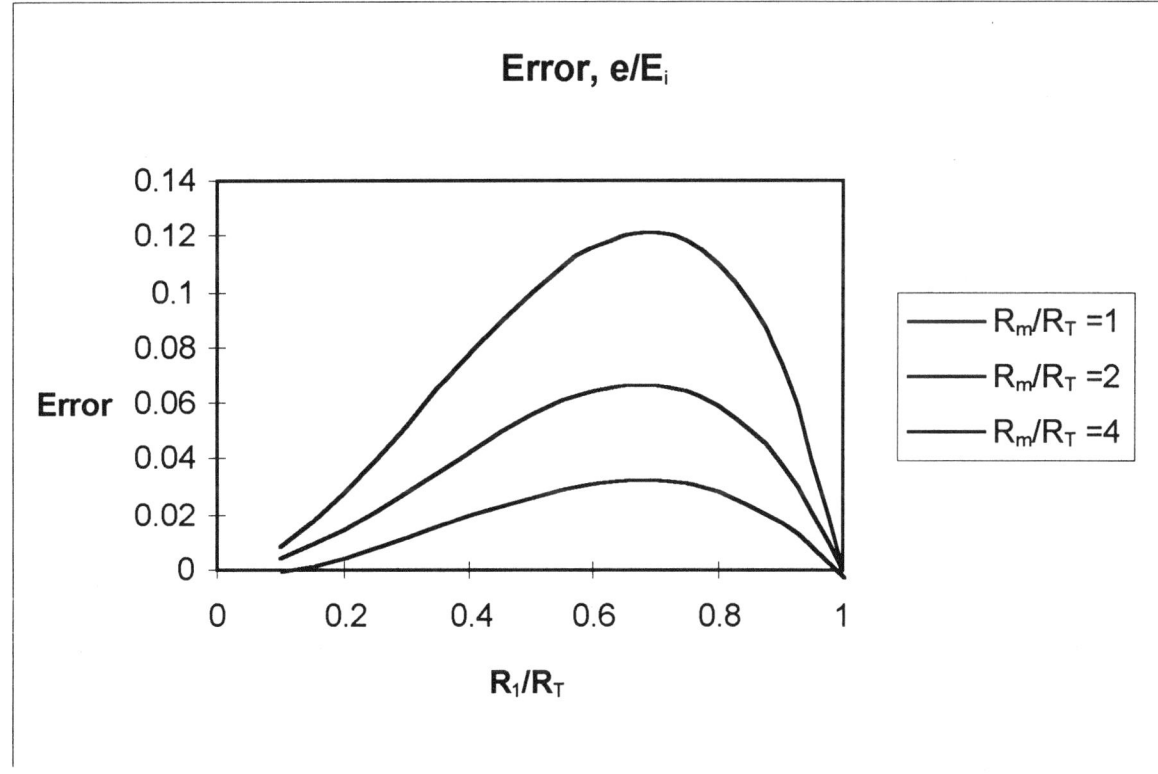

PROBLEM 6.3

KNOWN: $R_T = R_1 + R_2 = 500 \, \Omega$ $\quad R_m = 10,000 \, \Omega$

$R_1 = kR_T$ $\quad k = 0.5 \quad E_i = 10V$

FIND:
a) Loading error as a percentage of the output
b) Loading error as a percentage of the full scale output

SOLUTION:
The loading error may be defined as

$$e = E'_o - E_o$$

and in terms of a percentage of the output

$$\frac{E'_o - E_o}{E_o} = \frac{E'_o}{E_o} - 1$$

which yields, in terms of k

$$k\left[1 + \frac{(1-k)}{k}\left(k\frac{R_T}{R_m} + 1\right)\right] - 1$$

The loading error for this condition is 0.0125 or 1.25%.
As a percentage of full-scale output, E_i,

$$\frac{E'_o - E_o}{E_i} = k - \frac{k}{k + (1-k)\left(\frac{kR_T}{R_m} + 1\right)}$$

which produces a value for loading error of 0.62%. Since E_o is one-half of E_i, the loading errors are each 0.062 Volts.

PROBLEM 6.4

KNOWN:
$R_3 = R_4 = 200\ \Omega$
R_2 = variable resistor
$R_1 = 40x + 100\ \Omega$

FIND:
a) R_2 to balance the bridge with $x = 0$
b) Find a general expression for $x = f(R_2)$

ASSUMPTIONS: Zero Galvanometer error

SOLUTION: At balanced conditions

$$\frac{R_2}{R_1} = \frac{R_4}{R_3}$$

Thus $R_2 = (1)(100)$ and $R_2 = 100\ \Omega$
and in general $R_2 = 40x + 100$.

PROBLEM 6.5

KNOWN:
A Wheatstone bridge with $R_1 = 20x^2$ (x is a measured variable)
$R_3 = R_4 = 100\ \Omega$ $R_2 = 46\ \Omega$ at balanced conditions

FIND: x

SOLUTION: Since at balanced conditions

$$\frac{R_2}{R_1} = \frac{R_4}{R_3} = 1$$

Thus

$$\frac{46}{20x^2} = 1 \qquad x = \sqrt{\frac{46}{20}} = 1.52$$

PROBLEM 6.6

KNOWN: A sensor has a resistance of 500 Ω under conditions of no load, and a static sensitivity of 0.5 Ω/N. The bridge circuit has

$$R_1 = R_2 = R_3 = R_4 = 500 \, \Omega \quad \text{(initially)}$$

FIND: a) E_o for applied loads of 100, 200, and 350 N.

b) I_1

c) Repeat parts (a) and (b) with

$$R_m = 10 \, k\Omega$$

$$R_s = 500 \, \Omega$$

SOLUTION: For bridge in which all resistances are initally equal

$$\frac{E_o}{E_i} = \frac{\delta R / R}{4 + 2(\delta R / R)}$$

and with F_L = Load

$$\delta R = 0.5 F_L$$

$$\frac{E_o}{E_i} = \frac{0.5 F_L / 500}{4 + 2\left(\dfrac{0.5 F_L}{500}\right)}$$

This yields

F_L [N]	δR [Ω]	E_o [V]
100	50	0.238
200	100	0.455
350	150	0.652

The current flow through the sensor is I_1 and for an infinite meter resistance is

$$I_1 = \frac{E_i}{R_1 + R_2}$$

which yields

F_L [N]	R_1+R_2 [Ω]	I_1 [Amps]
100	1050	0.00952
200	1100	0.00909
350	1150	0.00870

c) Consider the circuit shown below, with values of δR equal to those for part (a)

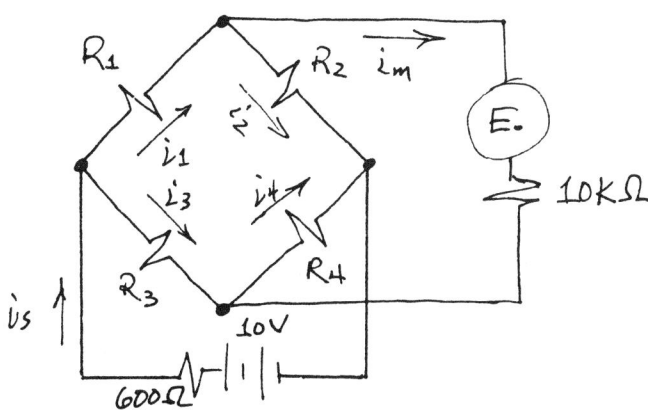

A circuit analysis of this bridge yields the following simultaneous equations governing the currents and potentials:

$E_i = I_s R_s + I_1(R_1+R_2) - I_m R_2$
$I_1 R_1 + I_m R_m - I_3 R_3 = 0$ $I_m R_m + I_4 R_4 - I_2 R_2 = 0$
$I_m + I_3 - I_4 = 0$ $I_s = I_1 + I_3$
$E_i = I_s R_s + I_3 R_3 + I_4 R_4$ $E_o = I_m R_m$

Solving these equations simultaneously yields

R_1 [Ω]	E_o [V]	I_1 [A]
550	0.104	0.0044
600	0.201	0.0042
650	0.291	0.0041

PROBLEM 6.7

KNOWN: Bridge circuit of Figure 6.36

FIND: Show that

$$C_2 = C_1\left(\frac{R_1}{R_2}\right)$$

is the requirement for a balanced bridge.

SOLUTION: With impedances (for an AC circuit)

$$Z_1 = R_1 \quad Z_2 = R_2 \quad Z_3 = \frac{1}{j\omega C_1} \quad Z_4 = \frac{1}{j\omega C_2}$$

from

$$E_o = E_i\left[\frac{Z_3}{Z_3 + Z_4} - \frac{Z_1}{Z_1 + Z_2}\right]$$

For this case

$$E_o = E_i\left[\frac{1}{1 + C_1/C_2} - \frac{1}{1 + R_1/R_2}\right]$$

which yields for $E_o = 0$

$$C_1 = C_2\left(\frac{R_1}{R_2}\right)$$

PROBLEM 6.8

KNOWN: A bridge circuit is to be used to calibrate a 500 Hz frequency source. The bridge resonance frequency is $f = \dfrac{1}{2\pi\sqrt{LC}}$

FIND: Bridge components L and C to yield a resonance frequency of 500 Hz.

SOLUTION: The product LC must be $(1000\pi)^2$ and
$$500 = \dfrac{1}{2\pi\sqrt{LC}}$$
which yields
$$LC = 9.87 \times 10^6 \text{ sec}^2$$

Combinations of L and C which yield the appropriate resonance frequency are shown below.

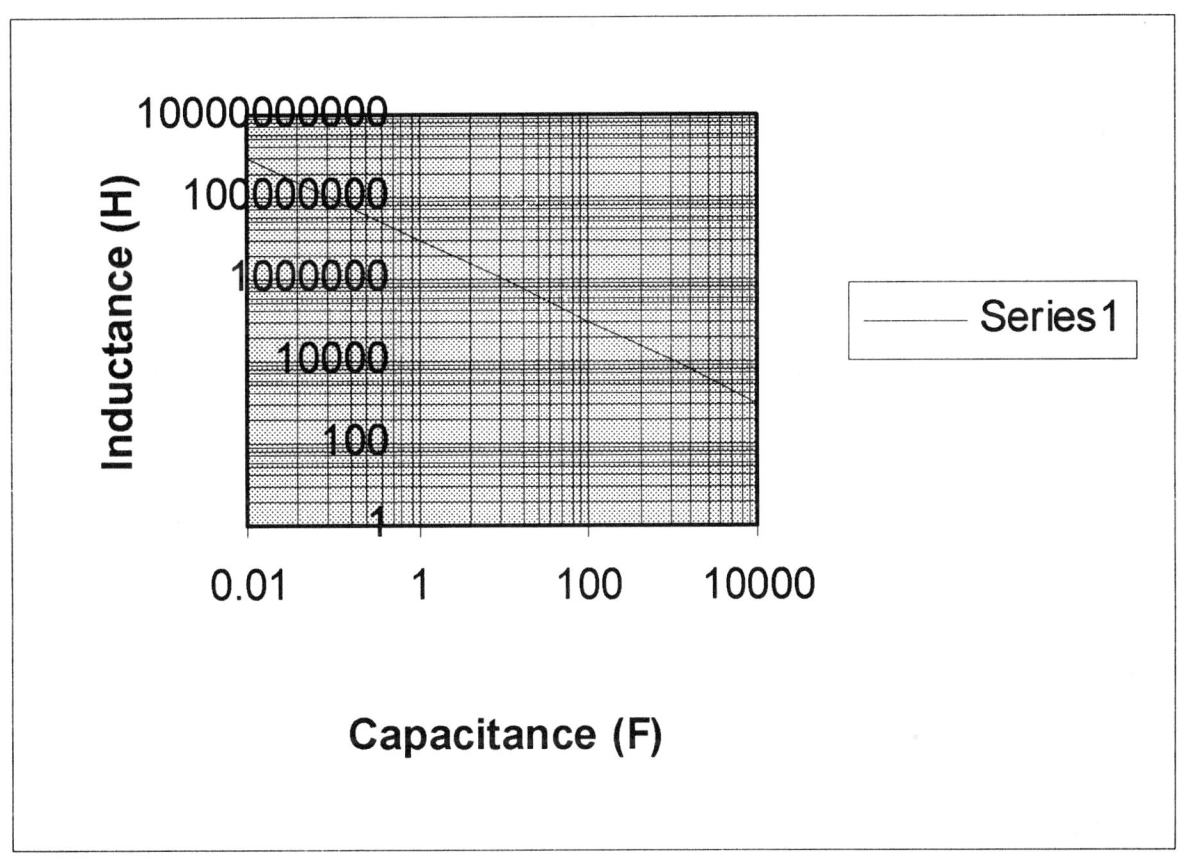

PROBLEM 6.9

KNOWN: A Wheatstone bridge wth

$R_1 = 200\ \Omega$ $\qquad$ $R_2 = 400\ \Omega$

$R_3 = 500\ \Omega$ $\qquad$ $R_4 = 600\ \Omega$ $\quad$ $E_i = 5$ V

FIND: a) E_o
b) E_o for $R_1 = 250\ \Omega$.

ASSUMPTIONS: The meter resistance R_g is infinite.

SOLUTION:

From (6.14)

$$E_o = E_i\left[\frac{R_1}{R_1 + R_2} - \frac{R_3}{R_3 + R_4}\right] = 5\left[\frac{200}{200 + 400} - \frac{500}{500 + 600}\right] = -0.606\ \text{V}$$

or with $R_1 = 250\ \Omega$

$$E_o = 5\left[\frac{250}{250 + 400} - \frac{500}{500 + 600}\right] = -0.35\ \text{V}$$

COMMENT: Clearly the bridge output is non-linear with R_1 over this range of values for R_1.

PROBLEM 6.10

KNOWN: A Wheatstone bridge having all resistances 500 Ω.

FIND: Plot the output voltage for :
  a) R_1 varies from 500 to 1000 Ω
  b) R_1 and R_2 change equally and in opposite directions between 500 and 600 Ω
  c) R_1 and R_3 change equally over the range 500 to 600 Ω

SOLUTION: From Equation 6.22

a)

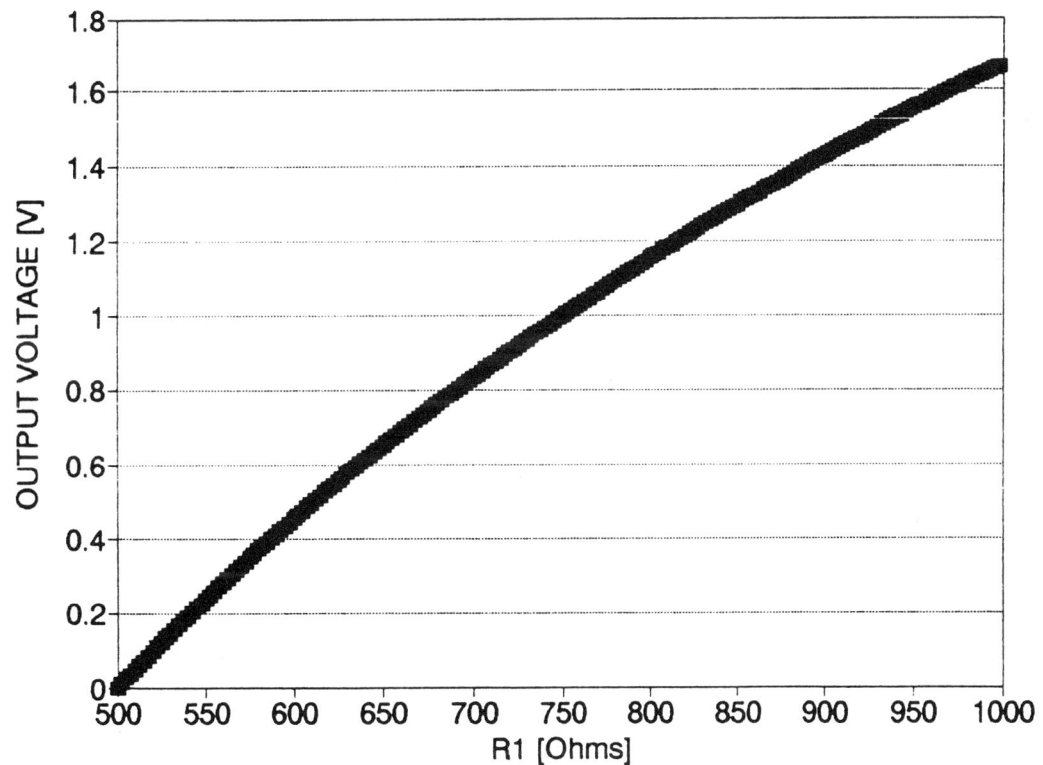

b)

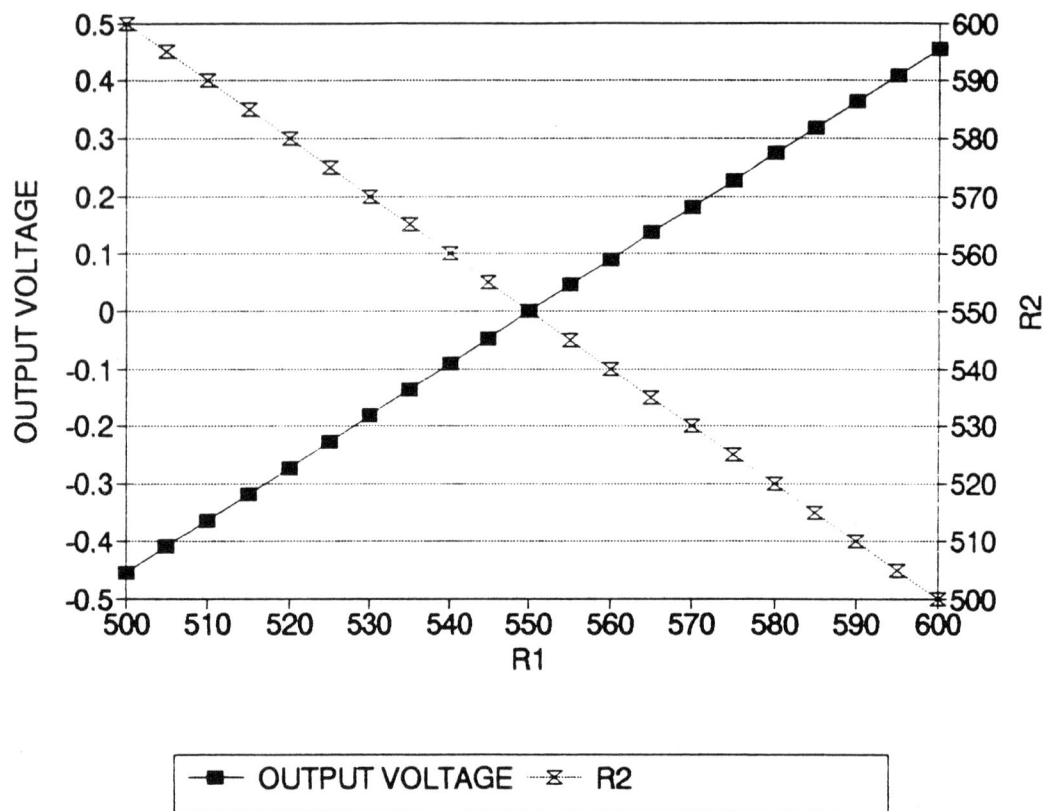

c) IDENTICALLY ZERO

PROBLEM 6.11

KNOWN: Potentiometer as shown in Figures 6.10 and 6.11.

$E_i = 10 \pm 0.1$ V $R_T = 100 \pm 1$ Ω

$R_g = 100$ Ω R_x = reading $\pm 2\%$

Null condition has negligible error.

FIND:

Design stage uncertainty in a measured value of voltage at nominal values of 2 and 8 V.

ASSUMPTIONS: Loading errors should be included in the analysis.

SOLUTION: From 6.35 (see Figure 6.16)

$$\frac{E_o}{E_i} = \frac{1}{1 + \frac{R_2}{R_1}\left(\frac{R_1}{R_g} + 1\right)}$$

Then at the design stage

$$u_{E_o} = \left[\sum\left(\frac{\partial E_o}{\partial x_i} u_{x_i}\right)^2\right]^{1/2}$$

For this case $E_o = f(E_i, R_x, R_g, R_T)$ and $R_T = R_1 + R_2$. At

$E_o = 2$ V $R_1 = 23.6$ Ω $R_2 = 76.4$ Ω
$E_o = 8$ V $R_1 = 88.3$ Ω $R_2 = 11.7$ Ω

The sensitivity indices may be evaluated as

$$\frac{\partial E_o}{\partial E_i} = \frac{E_i}{1 + \frac{R_2}{R_1}\left(\frac{R_1}{R_g} + 1\right)}$$

at 2 V $\frac{\partial E_o}{\partial E_i} = 2$, and at 8 V $\frac{\partial E_o}{\partial E_i} = 8$

$$\frac{\partial E_o}{\partial R_2} = \frac{-E_i/R_1 \left(\frac{R_1}{R_g}+1\right)}{\left[1+\frac{R_2}{R_1}\left(\frac{R_1}{R_g}+1\right)\right]^2}$$

AT 2V $\frac{\partial E_o}{\partial R_2} = 0.021$

AT 8V $\frac{\partial E_o}{\partial R_2} = 0.137$

$$\frac{\partial E_o}{\partial R_1} = \frac{E_i \left[-\frac{R_2}{R_1^2}\left(\frac{R_1}{R_g}+1\right)+\frac{R_2}{R_1}\left(\frac{1}{R_g}\right)\right]}{\left[1+\frac{R_2}{R_1}\left(\frac{R_1}{R_g}+1\right)\right]^2}$$

AT 2V $\frac{\partial E_o}{\partial R_1} = -0.055$

AT 8V $\frac{\partial E_o}{\partial R_1} = -0.00096$

$$\frac{\partial E_o}{\partial R_g} = \frac{\left(-R_2/R_g^2\right) E_i}{\left[1+\frac{R_2}{R_1}\left(\frac{R_1}{R_g}+1\right)\right]^2}$$

AT 2V $\frac{\partial E_o}{\partial R_g} = -0.0031$

AT 8V $\frac{\partial E_o}{\partial R_g} = -0.0075$

The uncertainty in R_2 is estimated as

$$R_2 = R_T - R_1$$

$$U_{R_2} = \left[U_{R_T}^2 + U_{R_1}^2\right]^{1/2}$$

AT 2V = 1.11 Ω

AT 8V = 2.03 Ω

At 2 V the uncertainty is estimated as

$$U_{E_o} = \left[(2 \cdot 0.1)^2 + (0.021 \cdot 1.11)^2 + (-0.055 \cdot 0.47)^2\right]^{1/2}$$

$$= \pm 0.203 \, V$$

and at 8 V as

$$U_{E_o} = \left[(8 \cdot 0.1)^2 + (0.137 \cdot 2.03)^2 + (0.00096 \cdot 1.8)^2\right]^{1/2}$$

$$= \pm 0.847 \, V$$

PROBLEM 6.12

KNOWN: Sinusoidal inputs having specified phase relations

FIND: Lissajous diagrams for the specified phase relations.

SOLUTION:

See figures below

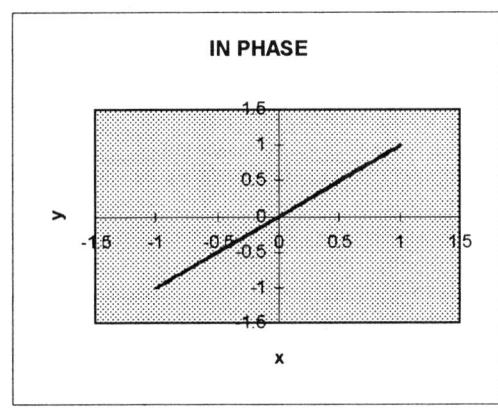

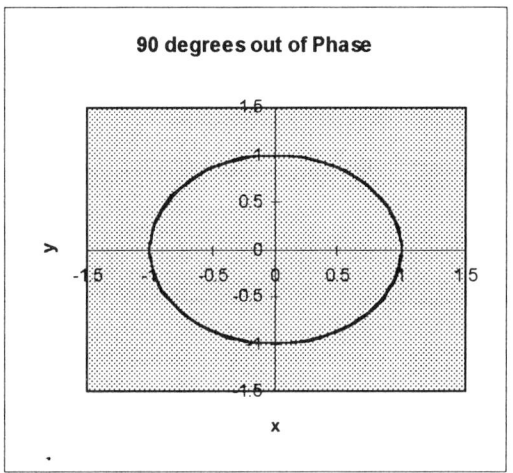

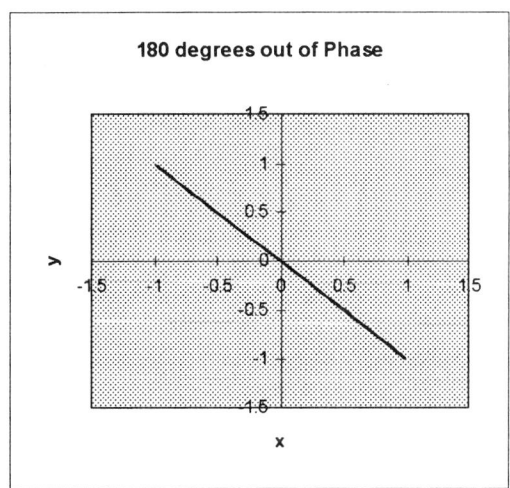

PROBLEM 6.13

KNOWN: $\sin\phi = \dfrac{y_i}{y_a}$

FIND: Show that phase relationships can be determined from a Lissajous diagram using the equation above.

SOLUTION:

Assume that two sine waves with a phase delay of ϕ are the input to the x and y terminals of an oscilloscope. Referring to Figure 6.38, the figure below shows y_a and y_i.

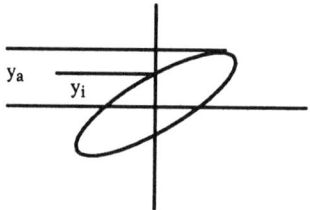

When $y = y_a$ then the y signal is a maximum, with $\sin\omega t = 1$, and with $y = A\sin\omega t$ and $y_a = A$. At the y-intercept, y_i where $x = 0$, $y_i = A\sin\phi$. Eliminating the amplitude A between these equations,

$$\frac{y_i}{y_a} = \frac{A\sin\phi}{A} = \sin\phi$$

PROBLEM 6.14

KNOWN: Phase lag occurs in electronic circuits.

FIND: Design an arrangement which uses Lissajous diagrams to determine phase lag.

SOLUTION:

The arrangement shown below will allow measurement of phase lag in an electronic circuit. See problem 6.15 for representative Lissajous diagrams.

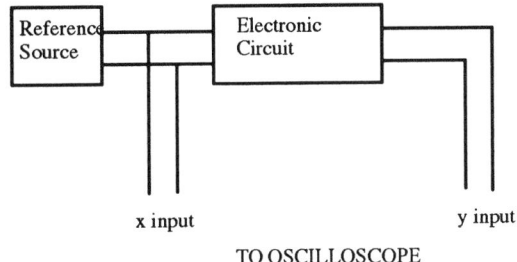

When $y = y_a$ then the y signal is a maximum, with $\sin \omega t = 1$, and with $y = A \sin \omega t$ and $y_a = A$. At the y-intercept, y_i where $x = 0$, $y_i = A \sin \phi$. Eliminating the amplitude A between these equations,

$$\frac{y_i}{y_a} = \frac{A \sin \phi}{A} = \sin \phi$$

PROBLEM 6.15

KNOWN: Two sinusoidal signals are to be compared using a dual trace oscilloscope. The frequency ratios are:

a) 1:1   b) 1:2   c) 2:1   d) 1:3   e) 2:3   f) 5:2

FIND: Construct appropriate Lissajous diagrams.

SOLUTION:
a)

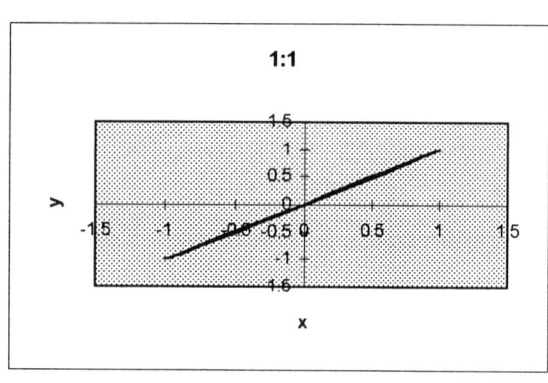

b)

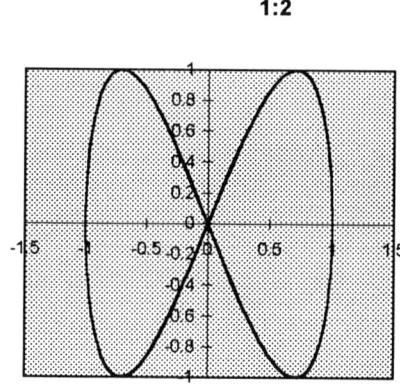

c)

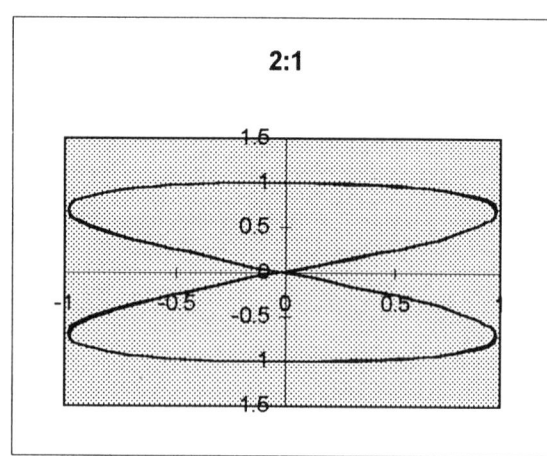

d)

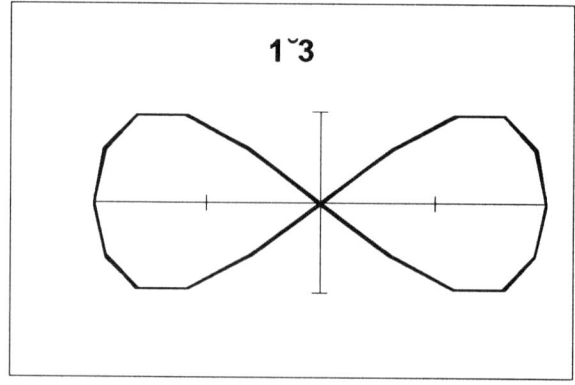

e)

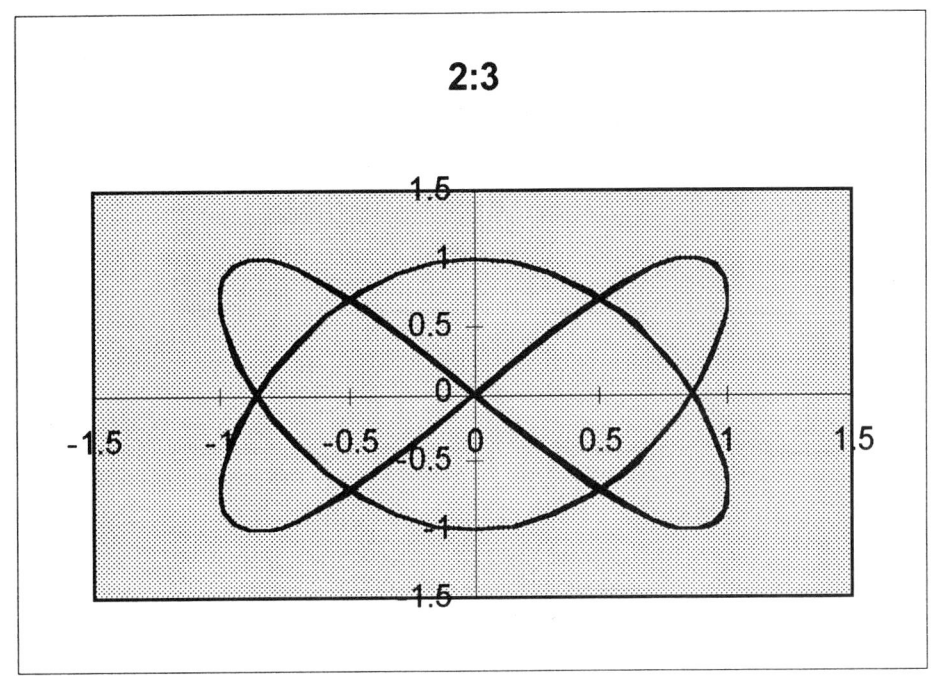

f)

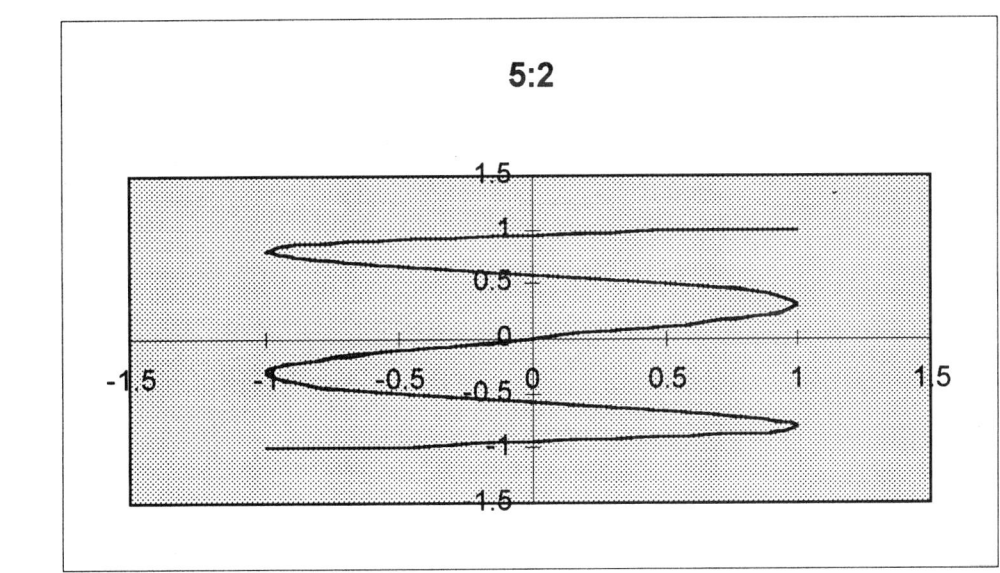

PROBLEM 6.16

KNOWN: RC filter: $k = 1$; $f_c = 100$ Hz

FIND: Attenuation at 10, 50, 75, and 200 Hz

ASSUMPTIONS: Filter is of the low pass, Butterworth type

SOLUTION

For a low pass RC Butterworth filter,

$$M(f) = 1/[1 + (f/f_c)^{2k}]^{0.5}$$

Recall that "attenuation" means reduction. In the context of a filter, a device which reduces ($M(f) < 1$) the output amplitude of targeted frequencies ($M(f) < 1$), it must correspond to a negative value of dynamic error. Dynamic error is given by

$$\delta(f) = M(f) - 1$$

Recall also that "gain" refers to a positive value of dynamic error.

For a low pass, RC Butterworth filter, $M(f) \leq 1$ always, which is consistent with a 1st order system behavior.

f [Hz]	M(f)	δ(f)	attenuation [%]
10	0.995	-0.005	0.5
50	0.894	-0.105	10.5
75	0.800	-0.200	20.0
200	0.447	-0.553	55.3

PROBLEM 6.17

KNOWN: Low-pass LC Bessel filter with k = 3
$f_c = 100$ Hz

FIND: Attenuation at 10, 50, 75, 200 Hz

SOLUTION

A three-stage, low-pass Bessel filter will have the transfer function (6.62)

$$G(s) = a_0/[a_0 + a_1 s + a_2 s^2 + a_3 s^3] = a_0/D_3(s)$$

For unit cut-off frequency, $D_3(s)$ has the form of (6.63), with k = 3

$$D_k(s) = D_3(s) = (2k-1)D_2(s) + s^2 D_1(s)$$

with $D_1(s) = s + 1$ and $D_2(s) = s^2 + 3s + 3$. The three stage, unit cut-off frequency transfer function is

$$G(s) = 15/[s^3 + 6s^2 + 15s + 15]$$

With $\omega_c = 2\pi f_c = 628$ rad/s, we substitute s/ω_c for s and solve for $s = j\omega$ to obtain

$$G(j\omega) = 15\omega_c^3/[-j\omega^3 - 6\omega^2 \omega_c + 15j\omega \omega_c^2 + 15\omega_c^3]$$

$$= M(\omega)e^{j\phi(\omega)}$$

From which, the magnitude

$$M(\omega) = 15\omega_c^3/[(15\omega\omega_c^2 - \omega_c^3)^2 + (15\omega_c^3 - 6\omega^2\omega_c)^2]^{0.5}$$

f	M(f)	Attenuation
10	1	0%
50	0.98	2%
75	0.94	6%
100	0.63	37%

PROBLEM 6.18

KNOWN: RC low-pass Butterworth filter
$-3 \text{ dB} \leq M(0 \leq f \leq 5 \text{ kHz})$
$M(f \geq 10 \text{ kHz}) < -30 \text{ dB}$

FIND: Values for R, C and k

SOLUTION

This problem is an open-ended design. One possible solution follows. For a low-pass, Butterworth filter,

$$M(f) = 1/[1 + (f/f_c)^{2k}]^{0.5}$$

where

$$f_c = 1/2\pi\tau \quad \text{and} \quad \tau = RC$$

We can set $f_c = 5$ kHz, such that $M(5 \text{ kHz}) = 0.707 = -3$ dB, and meet one constraint of the design. This gives,

$$\tau = RC = 1/2\pi(5000 \text{ Hz}) = 31.8 \ \mu s$$

One design combination could use $R = 3{,}180 \ \Omega$ and $C = 10$ pF.

The next step in the design is to determine the number of filter stages required to meet the attenuation constraint at 10 kHz. For at least 30 dB attenuation

$$-30 \text{ dB} \geq 20 \log M(10000 \text{ Hz})$$

then,

$$M(10000 \text{ Hz}) \leq 0.0316 = 1/[1 + (10 \text{ kHz}/5 \text{ kHz})^{2k}]^{0.5}$$

By trial and error, $k \geq 5$.

k	M(10000 Hz)
1	0.45
3	0.12
5	0.0312

PROBLEM 6.19

KNOWN: Circuit of Figure 6.39.

FIND: Thevenin equivalent.

In open-circuit operation,

$$E_o = I_n R_n$$

In short-circuit operation,

$$I_n = E_{th}/R_{th}$$

or,

$$E_{th} = I_n R_n$$

and

$$R_{th} = E_{th}/I_n$$

PROBLEM 6.20

KNOWN: $Z_1 = 1 \times 10^9 \Omega$

FIND: Z_m required to keep $E_I/E \leq 0.1\%$

SOLUTION

For the circuit in Figure 6.40, (6.38) can be written:

$$E_m = E_1 Z_m/(Z_m + Z_1)$$

where E_1 is the Thevenin equivalent voltage. For $Z_1 = 10^9 \Omega$

$$E_m/E_1 = Z_m/(Z_m + 10^9)$$

Then the relative loading error or signal attentuation is

$$e_I/E_1 = 1 - E_m/E_1 \qquad (6.39)$$

Z_m [Ω]	$(1 - E_m/E_1)\%$
10^6	99.9%
10^9	50%
10^{12}	0.1%

Z_m must be at least $10^{12}\Omega$ to meet the constraint. This requires $Z_m/Z_1 = 10^{12}/10^9 \geq 1000$.

COMMENT

Loading error is only a problem if it goes undetected. It can be prevented by proper design as it is easily predicted.

PROBLEM 6.21

KNOWN: Circuit of Figure 6.41

FIND: E_o/E_i for circuit.

SOLUTION

Across the divider, (6.8) gives:

$$E_o = E_i(60)/160 = 0.375 E_i \quad \text{or} \quad (E_o/E_i)_{divider} = 0.375$$

This is equivalent to a gain of

$$G\,[d\beta] = 20 \log (E_o/E_i) = -8.5\; d\beta$$

The overall circuit gain based on $G_{amp} = +32 d\beta$, $G_{filter} = -12\; d\beta$, and the divider circuit is:

$$E_o/E_i = (32 - 12 - 8.5)\; d\beta = 11.5\; d\beta = 3.76$$

PROBLEM 6.22

KNOWN: Temperature measurement system employing a resistance temperature detector, and a Wheatstone bridge circuit having the following characteristics.

$$R = R_o\left[1+\alpha(T-T_o)\right] \quad R_o = 100 \: \Omega \text{ at } 0°C \quad R_3 = R_4 = 500 \: \Omega$$

There are two error sources, the sensitivity of the galvanometer, and the accuracy of the fixed resistors.

FIND: Design a combination of uncertainty in the fixed resistors and the current flow through the galvanometer to provide an uncertainty in temperature of $\pm 1°C$.

SOLUTION:

The sensitivity indices are needed to determine the required accuracy of the detector resistance measurement to yield an uncertainty of $\pm 1°C$, and the uncertainty of the measurement of R_1 in the bridge. The required level of uncertainty in the resistance measurement can be determined from

$$\frac{\partial R}{\partial T} = \frac{\partial}{\partial T}\left[R_o\left[(1+\alpha(T-T_o))\right]\right] = \alpha R_o$$

Knowing that $u_T = \pm 1°C$, the required uncertainty in R_1 is found from

$$u_R = \frac{\partial R}{\partial T}u_T = \alpha R_o u_T = (0.00395)(100)(1) = 0.395 \: \Omega \qquad (1)$$

Then to evaluate the sensitivity of R_1 to the uncertainty in I_g, sequential pertubation can be employed with equation 6.22, to yield a value of 0.000045 A/Ω, or 22.2 Ω/mA. The uncertainty in R_1 may be expressed

$$u_{R_1} = \sqrt{\left(\frac{\partial R_1}{\partial I_g}u_{I_g}\right)^2 + \left(\frac{\partial R_1}{\partial R_2}u_{R_2}\right)^2 + \left(\frac{\partial R_1}{\partial R_3}u_{R_3}\right)^2 + \left(\frac{\partial R_1}{\partial R_4}u_{R_4}\right)^2} \qquad (2)$$

where

$$\frac{\partial R_1}{\partial R_2} = \frac{R_3}{R_4} \quad \frac{\partial R_1}{\partial R_3} = \frac{R_2}{R_4} \quad \frac{\partial R_1}{\partial R_4} = \frac{-R_2 R_3}{R_4^2}$$

With appropriate assumptions of uncertainty in R's and I_g the value of u_T can be evaluated, and an iterative process used to satisfy the constraint.

PROBLEM 7.1

KNOWN: $E(t) = 5\sin 2\pi t$ mV

FIND: Convert to a discrete time series and plot

SOLUTION

The signal is converted to a discrete time series for sample time increments of 0.125, 0.30, and 0.75 s and plotted below. The time increment of 0.75 s fails the Sampling theorem criterion and the effect can be seen as an apparent change in the signal frequency content.

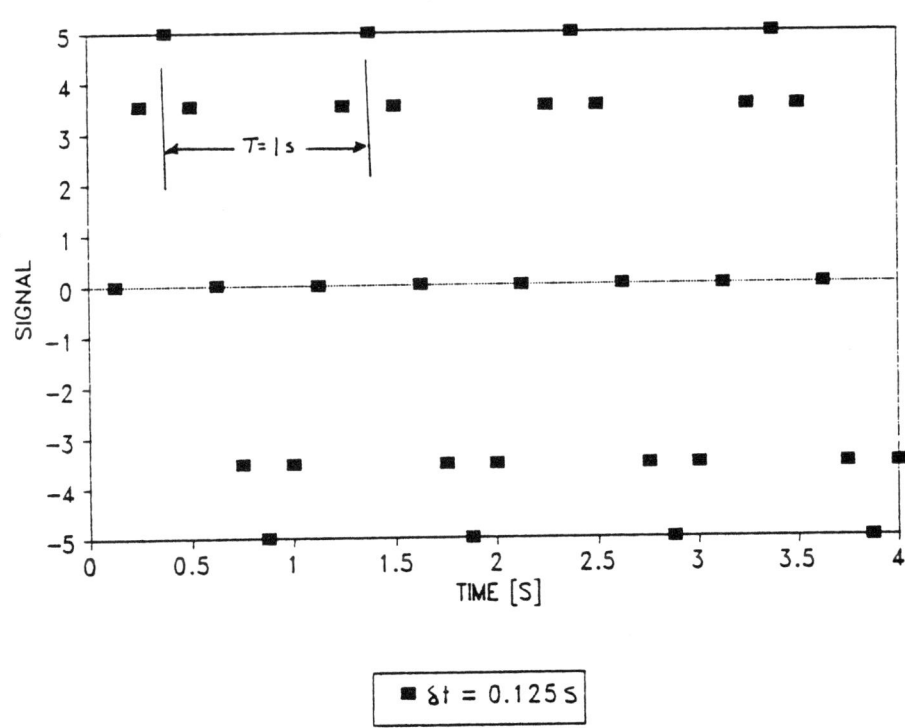

7.1

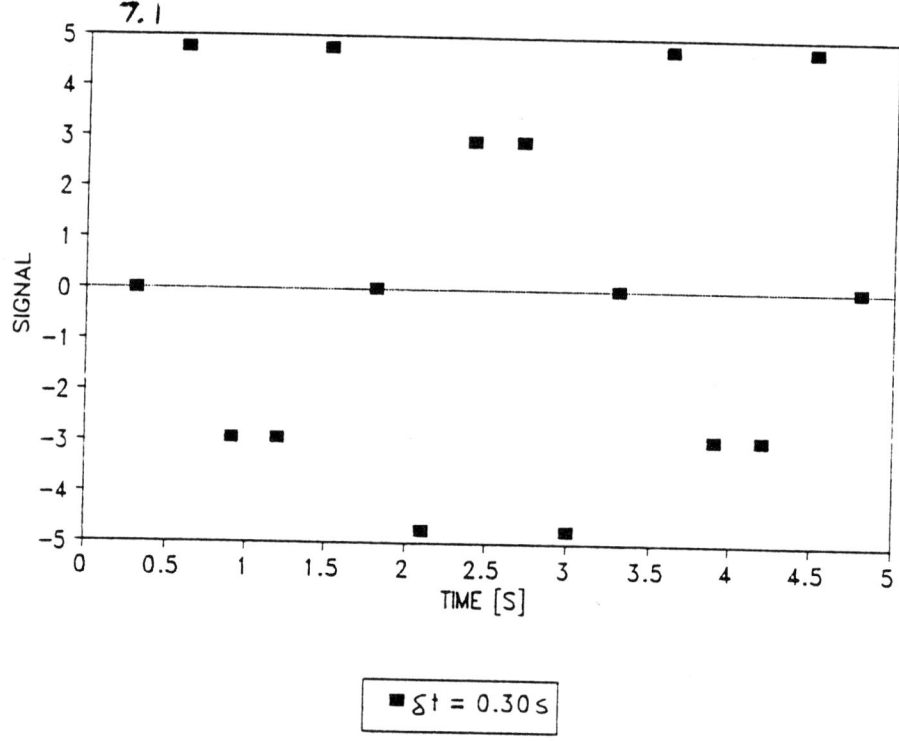

■ δt = 0.30 s

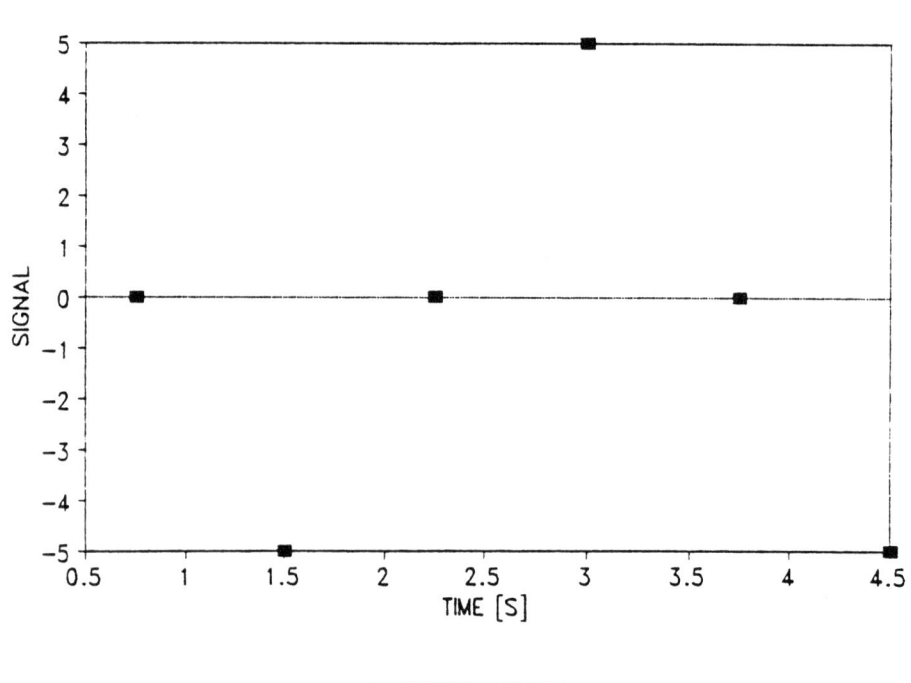

■ δt = 0.75 s

PROBLEM 7.2

KNOWN: Repeat Problem 7.1 using N = 128 points.

FIND: The discrete Fourier transform for each series.

SOLUTION

The DFT for the time series representations of E(t) using N = 128 and δt = 0.125, 0.30, and 0.75 s, respectively was performed using the DFT algorithm in Appendix A (any such algorithm will work) and shown below. The DFT returns an exact Fourier transform of the discrete time series. However, whether this DFT exactly represents E(t) depends on the criteria:

(1) $f_s = 1/\delta t > 2f$
(2) $m/f_1 = N\delta t \quad m = 1,2,3,...$

With f = 1 Hz and f_1 = f: (a) δt = 0.125 s and N = 128:

$$f_s = 1/0.125s = 8 \text{ Hz} > 2 \text{ Hz}$$

m/1 Hz = 128/0.125 or m = 16 an exact integer value.

Another way to look at this second criterion: the DFT resolution $\delta f = 1/N\delta t$ = 0.0625 Hz, of which 1 Hz is an exact multiple.

Both criteria are met. Therefore, this DFT will exactly represent E(t) in both frequency and amplitude, as shown below.

(b) δt = 0.3 s and N = 128

$$f_s = 1/0.3s = 3.3 \text{ Hz} > 2 \text{ Hz}$$

m/1 Hz = 128/0.3 or m = 38.4 not an exact integer value.

Criterion (2) is not met. Therefore, this DFT will not exactly represent E(t) in amplitude and spectral leakage will occur, as seen below.

(c) $\delta t = 0.75$ s and $N = 128$

$$f_s = 1/0.75s = 1.33 \text{ Hz} < 2 \text{ Hz}$$

$$m/1 \text{ Hz} = 128/0.75 \quad \text{or} \quad m = 96 \quad \text{an exact integer value.}$$

Criterion (1) is not met. Therefore, this DFT will not exactly represent $E(t)$ in frequency. As seen below, an alias frequency at 3Hz appears in the DFT.

COMMENT

Signal reconstruction from the DFT can be performed within the DFT algorithm using the spectral amplitudes computed. These are shown below.

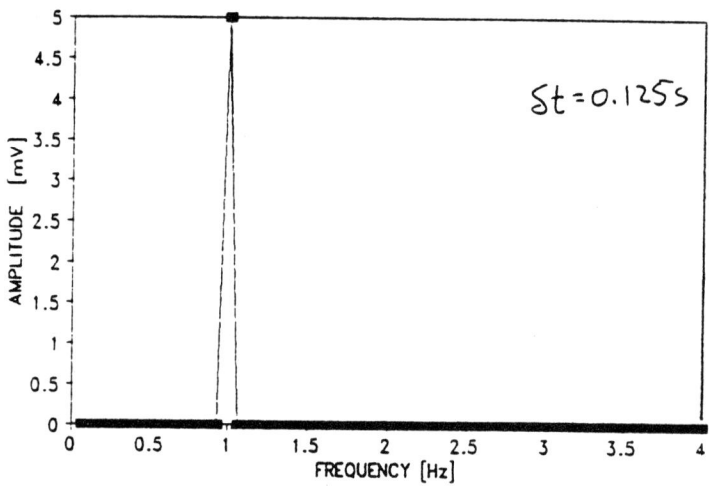

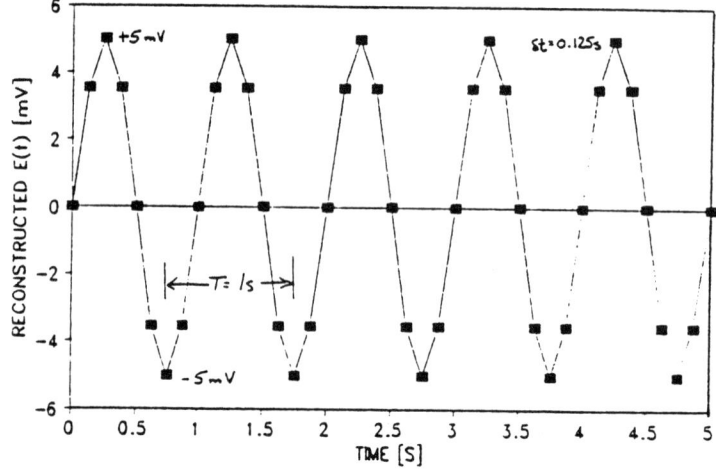

7.2

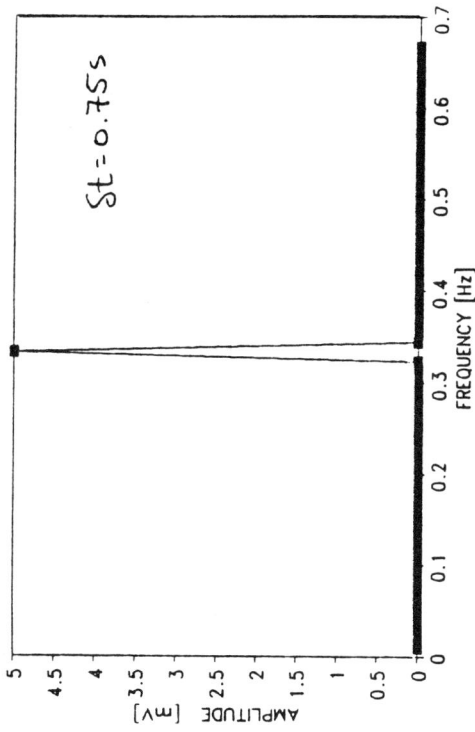
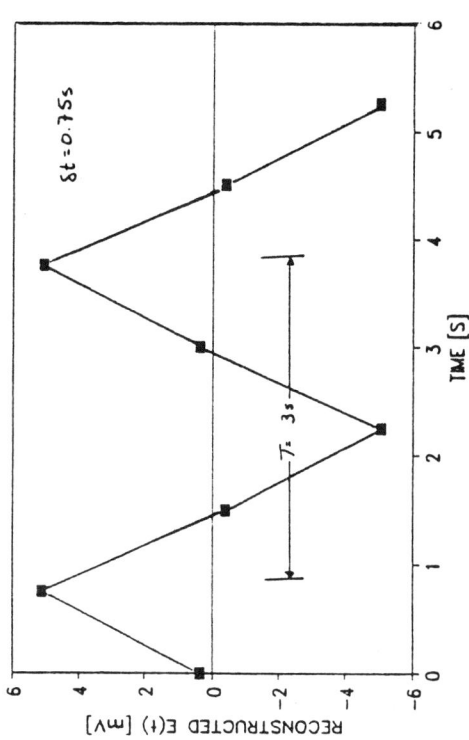
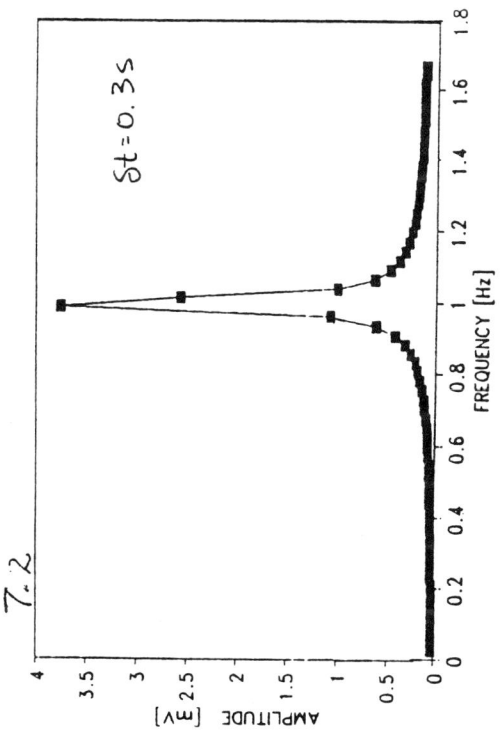
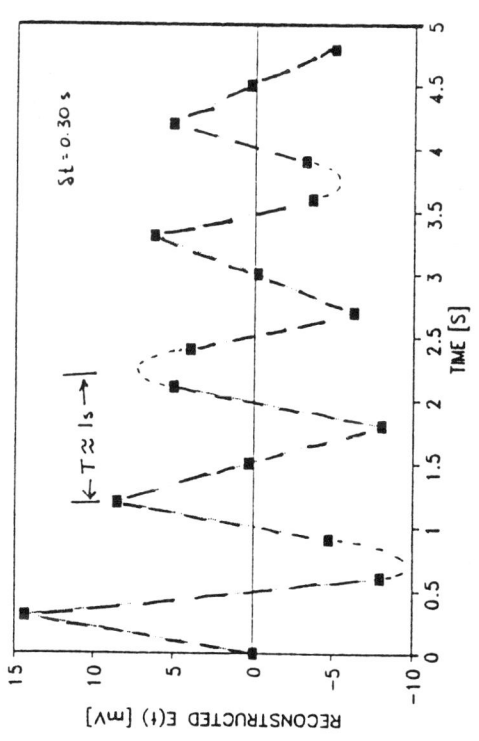

PROBLEM 7.3

KNOWN: $T(t) = 2 \sin 4\pi t$ °C
$f_s = 4$ and 8 Hz; $N = 128$

FIND: Compute the Fourier transform from the resulting series.

SOLUTION

The fundamental frequency of this signal is $f_1 = 2$ Hz. From the Sampling theorem, an appropriate sample rate is
$f_s > 2f_1$ or $f_s > 4$Hz At $f_s = 4$ Hz the Sampling theorem is not maintained whereas at $f_s = 8$ Hz the Sampling theorem is maintained. For $f_s = 4$ Hz, see the COMMENT below.

For amplitude fidelity, $mT_1 = N\delta t$ which can be written as $m/f_1 = N/f_s$ $m = 1, 2, ...$ With $f_s = 8$ Hz and $N = 128$, $m = Nf_1/f_s = N\delta t/T_1 = 8$ an exact integer. We can expect that the resulting DFT will be an exact representation of $T(t)$, as seen below.

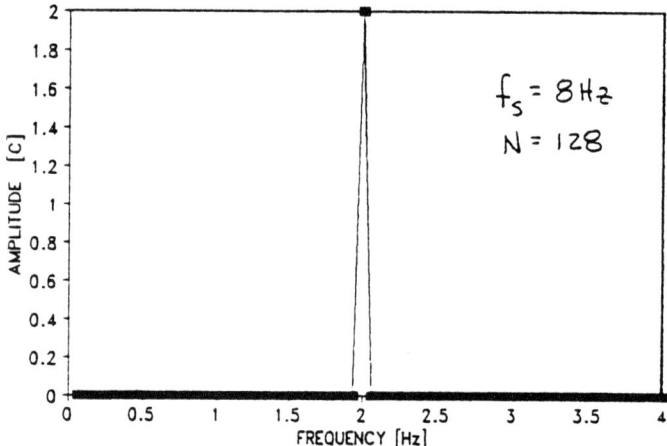

COMMENT

The result of sampling at $f_s = 2f$ is somewhat interesting. A non-unique reconstruction of $T(t)$ can result which will depend wholly on the initial condition of $T(t)$ at the time sampling commences. It is because of this dilemma that some texts will cite erroneously the Sampling Theorem criterion as $f_s \geq 2f$.

PROBLEM 7.4

KNOWN: f and f_s

FIND: The alias frequency, f_a

SOLUTION

The Nyquist or folding frequency is defined as $f_N = f_s/2$. All frequencies in the sampled signal that are above f_N will be folded back to lower frequencies below f_N, a process depicted in Figure 7.3.

(a) f = 50 Hz f_s = 90 Hz

$$f_N = f_s/2 = 45 \text{ Hz.}$$

The ratio, f/f_N = 1.11, that is f = 1.11f_N. Referring to Figure 7.3, a frequency of 1.11f_N will be folded back to a frequency of 0.889f_N.

$$f_a = 0.889 f_N = 40 \text{ Hz}$$

(b) f = 1.5kHz f_s = 2kHz

$$f_N = f_s/2 = 1\text{kHz.}$$

The ratio, f/f_N = 1.5, that is f = 1.5f_N. Referring to Figure 7.3, a frequency of 1.5f_N will be folded back to a frequency of 0.5f_N.

$$f_a = 0.5 f_N = 500 \text{ Hz}$$

(c) f = 8 Hz f_s = 6 Hz

$$f_N = f_s/2 = 3 \text{ Hz.}$$

The ratio, f/f_N = 8/3, that is f = 8f_N/3. Referring to Figure 7.3, a frequency of 8f_N/3 will be folded back to a frequency of 2f_N/3.

$$f_a = 2f_N/3 = 2 \text{ Hz}$$

(d) $f = 21$ Hz $f_S = 8$ Hz

$$f_N = f_S/2 = 4 \text{ Hz}.$$

The ratio, $f/f_N = 5.25$, that is $f = 5.25f_N$. Referring to Figure 7.3 and extrapolating out, a frequency of $5.25f_N$ will be folded back to a frequency of $0.75f_N$.

$$f_a = 0.75f_N = 3 \text{ Hz}.$$

PROBLEM 7.5

KNOWN: $E(t) = \sin 2\pi t + 2 \sin 8\pi t$
$f_s = 16$ Hz

FIND: Fourier transform of a discrete time series representation of $E(t)$.
Reconstruct $E(t)$ from the Fourier transform.

SOLUTION

Begin by generating a discrete time series. We can rewrite $E(t)$ as

$$E(t) = \sin 2\pi f_1 t + 2 \sin 2\pi f_2 t$$

with $f_1 = 1$ Hz and $f_2 = 4f_1$. An exact DFT representation of the $E(t)$ will result if the discrete time series is created using

(1) $f_s > 2f_2$
(2) $mT_1 = N\delta t$ or $m/f_1 = N/f_s$ $m = 1,2,...$

For (1) to be true, $f_s > 8$ Hz. This constraint is met, so frequency will be correctly represented. For (2) to be true, we need

$$N = mT_1/\delta t = mf_s/f_1 = m(16 \text{ Hz})/1 \text{ Hz} = 16m$$

That is for amplitude to be correctly represented, N must be an exact multiple of 16 (e.g. 16, 32, ...). We use $N = 32$ to construct the discrete time series. The time series and its DFT are computed using the DFT algorithm in Appendix A (see Instructor's Note, Problem Solution 6.16). The time signal is then reconstructed from the spectral information by

$$E(t) = \Sigma\, C_k \sin(2\pi t + \tan^{-1} A_k/B_k)$$

The DFT and reconstructed time signal are shown below.

7.5

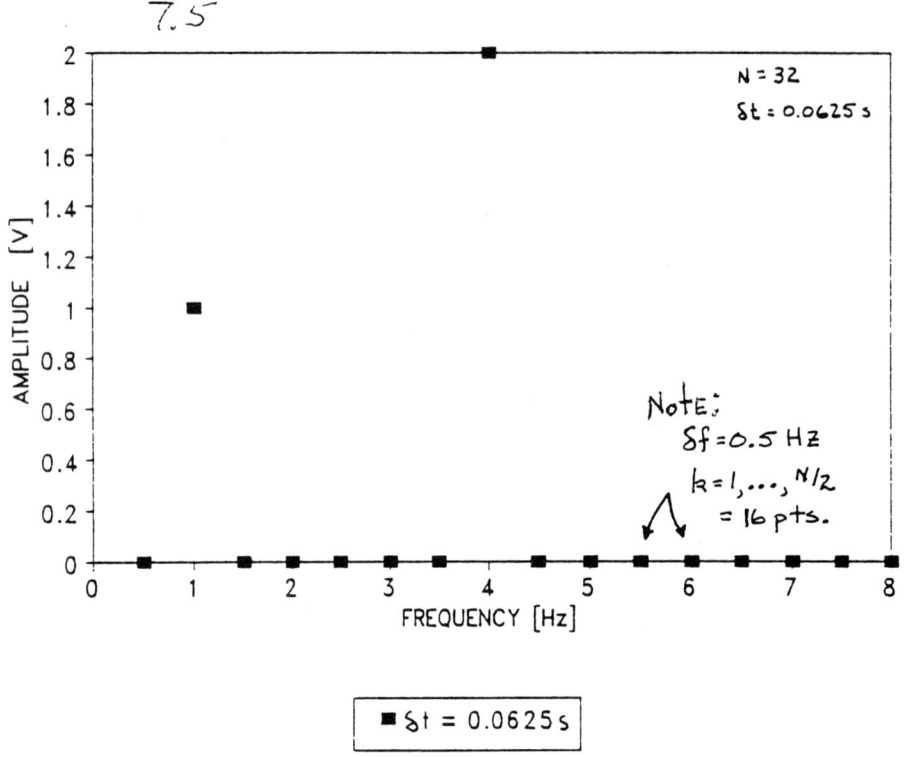

N = 32
δt = 0.0625 s

NOTE:
δf = 0.5 Hz
k = 1,..., N/2
= 16 pts.

■ δt = 0.0625 s

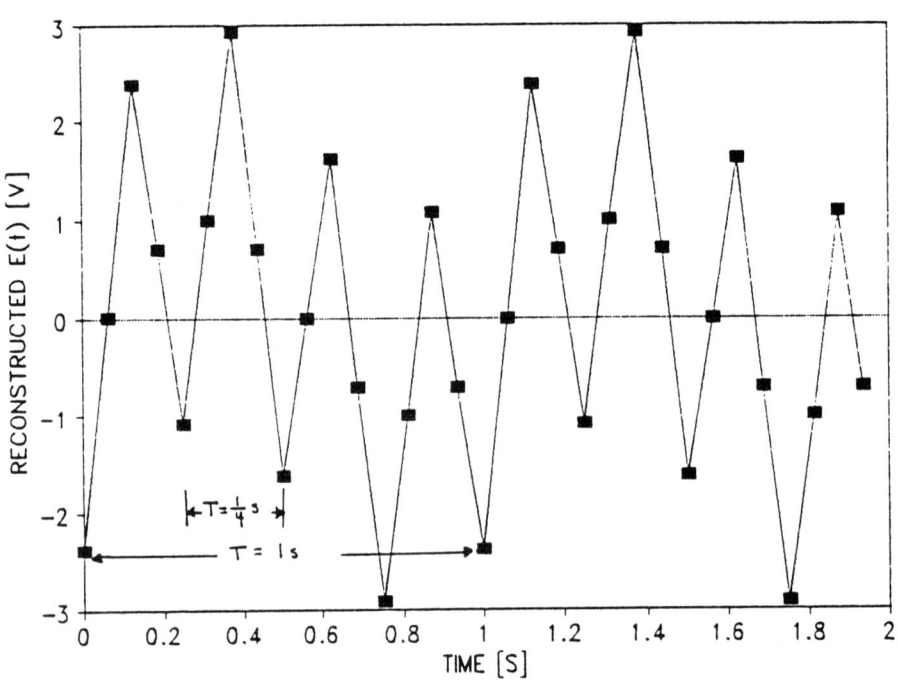

|← T = 1/4 s →|
← T = 1 s →

PROBLEM 7.6

KNOWN: $y(t) = \sum_n (4/(2n-1)\pi) \sin[2\pi(2n-1)t/10]$

FIND: Appropriate values for f_s and $N\delta t$ with $f_m \leq 2$ Hz and $N = 2^M$.

SOLUTION

Filtering the signal at and below 2 Hz limits this series representation of $y(t)$ to 5 terms. Each nth term of the series has the frequency $f_n = (2n-1)/10$. The fundamental frequency ($n=1$) is $f_1 = 0.1$ Hz.

In order to properly construct a discrete time series, the following criteria should be met:

(1) $f_s > 2f_m$ Sampling Theorem for frequency content

(2) $mT_1 = N\delta t$ $m = 1,2,...$ Amplitude content

where $T_1 = 1/f_1$. Because $f_m = 2$ Hz, criterion (1) is met by setting $f_s > 4$ Hz.

There are many different correct solutions to this problem. One such solution is to set $f_s = 8$ Hz. Then this requires,

$$m(10 \text{ s}) = N(1/8 \text{ Hz}) \quad m = 1,2,...$$

Also, we must set $N = 2^M$ where M is an integer. Criterion (2) is met with $N = 2^{10} = 1024$. Hence,

$$f_s = 8 \text{ Hz} \quad N = 1024 \quad \text{and} \quad N\delta t = N/f_s = 1024/8 = 128 \text{ s}$$

COMMENT

It should be stressed that there are usually many combinations of f_s and $N\delta t$ that will meet the criteria of (1) and (2) in a given problem. In a practical problem, additional restrictions, such as the maximum sample rate available or the maximum data set size that can be stored, will limit the number of possible combinations. Filter selction may affect the choice of f_s. Some later problems in this Chapter will impose such limitations. For example, problem 7.25 limits f_s and discusses the relationship between f_s and filter stage requirements.

PROBLEM 7.7

KNOWN: Straight Binary Number

FIND: Equivalent base 10 representation

SOLUTION

(a) $1010 = 1*2^3 + 0*2^2 + 1*2^1 + 0*2^0 = 10_{10}$

(b) $11111 = 1*2^4 + 1*2^3 + 1*2^2 + 1*2^1 + 1*2^0 = 31_{10}$

(c) $10111011 = 1*2^7 + 0*2^6 + 1*2^5 + 1*2^4 + 1*2^3 + 0*2^2 + 1*2^1 + 1*2^0 = 187_{10}$

(d) $1100001 = 1*2^6 + 1*2^5 + 0*2^4 + 0*2^3 + 0*2^2 + 0*2^1 + 1*2^0 = 97_{10}$

PROBLEM 7.8

KNOWN: One complementary binary number

FIND: Equivalent base 10 representation

SOLUTION

In one complementary binary, the value of the most significant bit (MSB) assigns the sign of the number (see p. 280):

(a) $0111 = 0*(-1) + 1*2^2 + 1*2^1 + 1*2^0 = +7_{10}$

(b) $1001 = 1*(-1) + 0*2^2 + 0*2^1 + 1*2^0 = -1_{10}$

(c) $01111111 = 0*(-1) + 1*2^6 + 1*2^5 + 1*2^4 + 1*2^3 + 1*2^2$
$\qquad + 1*2^1 + 1*2^0 = +127_{10}$

(d) $11111111 = 1*(-1) + 1*2^6 + 1*2^5 + 1*2^4 + 1*2^3 + 1*2^2$
$\qquad + 1*2^1 + 1*2^0 = -127_{10}$

PROBLEM 7.9

FIND: Equivalent two's complement representation

SOLUTION

For example, assuming a 12 bit binary number, two's complement assigns a 0 to the MSB is (+) with succesive bits in straight binary (0 HIGH, 1 LOW), or a 1 if (-) with successive bits complementary (1 HIGH, 0 LOW). Because two's complement has only a single zero add 1 to its count (negative number).

(a) 10_{10} = 0000 0000 1010

(b) -10_{10} = 1111 1111 0110

(c) -247_{10} = 1111 0000 1000

(d) 1013_{10} = 0011 1111 0101

PROBLEM 7.10

KNOWN: Two's complement code; M = 8

SOLUTION

Two's complement code uses the MSB as the sign bit, with a 0 for a positive number, and straight binary. The largest positive number is:

$$0111\ 1111 = (+) + 64 + 32 + 16 + 8 + 4 + 2 + 1 = 127_{10}$$

Adding a one to this binary number yields a negative number. Negative numbers use complementary binary (0 = HIGH, 1 = LOW). Note: Because two's complement has only a single zero add 1 to a negative count.

$$1000\ 0000 = (-1)128 = -128_{10}$$

PROBLEM 7.11

KNOWN: Two's complement code with M = 8

SOLUTION

Two's complement code uses the MSB as the sign bit, with a 1 for a negative number, and complementary binary (0 = HIGH, 1 = LOW). Note: Because two's complement has only a single zero add 1 its negative count. The largest negative number is:

$$1000\ 0000 = (-1) + 64 + 32 + 16 + 8 + 4 + 2 + 1 = -128_{10}$$

Subtracting one from this (the binary not the decimal) number:

$$0111\ 1111 = (+1)127 = +127_{10}$$

PROBLEM 7.12

KNOWN: Dual-slope integration A/D

FIND: List error sources. Derive a relationship between the uncertainty in the digital result and the slope of the integration process.

SOLUTION

Possible error sources include (see Figure 7.11): 1. errors in controlling t_1; 2. errors in measuring t_m; 3. errors in applied E_{ref}

Since the slope is proportional to E_{ref}: $E_i = E_{ref}(t_m/t_1)$. The uncertainty in the measured voltage can be expressed as:

$$u_{E_i} = \pm \left[\left(\frac{\partial E_i}{\partial E_{ref}} u_{ref} \right)^2 + \left(\frac{\partial E_i}{\partial (t_m/t_1)} u_{t_m/t_1} \right)^2 \right]^{1/2}$$

The value of E_{ref}, which determines the slope, then contributes to the uncertainty in two ways: 1. The uncertainty in E_{ref} will vary linearly with t_m. 2. Because the sensitivity index for the uncertainty in t_m/t_1 equals E_{ref}, the slope directly affects the sensitivity of the measured voltage through the uncertainty in time determination.

The devices that contribute to these uncertainties are associated with the frequency source and the counter.

PROBLEM 7.13

KNOWN: A/D Converter
$M = 4, 8, 12, 16$
-5 to 5 V bipolar

FIND: Q, dynamic range

SOLUTION

The resolution of an M-bit A/D converter is expressed as

$$Q = E_{fsr}/2^M \quad (7.13)$$

For this problem, $E_{fsr} = 10$ V (i.e. -5 to 5 V)

The dynamic range can be expressed in terms of the signal-to-noise ratio (SNR),

$$SNR\,[dB] = 20 \log 2^M \quad (7.14)$$

M	Q [V]	SNR [dB]
4	0.62500	-24
8	0.03906	-48
12	0.00244	-72
16	0.00015	-96

PROBLEM 7.14

KNOWN: A/D Converter: $M = 12$; $E_{FSR} = 5V$; $e_1/E_{FSR} = 0.0003$

FIND: e_Q, e_{max}, u_E/E

SOLUTION

$$Q = 5V/2^{12} = 1.2 \text{ mV}$$

$$e_Q = \pm Q/2 = \pm 0.6 \text{ mV}$$

The maximum error that could be present in any measurement is found by summing all known errors:

$$u_{max} = e_{max} = e_Q + (E_{FSR})(e_1/E_{FSR}) = 0.6 + 1.5 = 2.1 \text{ mV}$$

However, the probable error provides a reasonable estimate: (more)

$$u_E = \pm (e_Q^2 + e_1^2)^{.5} = \pm(.6^2 + 1.2^2)^{.5} = 1.6 \text{ mV} \quad (95\%)$$

or $u_E/E = 0.032\%$.

COMMENT

The maximum error assumes that all possible errors are at their maximum value during a measurement.

PROBLEM 7.15

KNOWN: Single ramp A/D converter: $M = 8$; $E_{FSR} = 10$ V; $f_{clock} = 2.5$ MHz
Comparator: $E_{th} = 1$ mV
$E_i = 6.000$ V. $E_i = 6.035$ V.

FIND: Binary output. Conversion times. Resolution.

SOLUTION

Resolution: $Q = 10 \text{ V}/2^8 = 39$ mV (or 0.4% FSO) for a ramp slope equivalent to 0.039 V/time step. For $E_i = 6.000$ V, the number of steps required is:

$$E_i/Q = 153.6 \pm 154 \text{ steps}$$

With a single count for each step, $154_{10} = 1001\ 1010$ in straight binary.

Repeating for $E_i = 6.035$ V:

$$E_i/Q = 155 \text{ steps or } 155_{10} = 1001\ 1011 \text{ in straight binary.}$$

With $f_{clock} = 2.5$ MHz, each step requires 0.4 μs. So the conversion times are:

$$E_i = 6.000 \text{ V}: \quad t = (0.4 \mu s)(154) = 102.4\ \mu s$$

The maximum conversion time required for a given input occurs at its maximum count or 2^8: $t_{max} = (0.4\ \mu s)(256) = 102.4\ \mu s$ The average: $t_{avr} = (0.4\ \mu s)(128) = 51.2\ \mu s$

PROBLEM 7.16

KNOWN: A/D converter: $M = 10$; $E_{FSR} = 10$ V. $f_{clock} = 1$ MHz.

FIND: Compare conversion times for successive approximation versus dual-ramp operation.

SOLUTION

Successive approximation Method: A maximum conversion time would result after M trials. For $M = 10$,

$$t_{max} = M/f_{clock} = 10/1 \text{ MHz} = 10 \text{ }\mu s.$$

Dual-ramp methods: A maximum conversion time results after $(2)(2^M) = (2^{M+1})$ trials. For $M = 10$,

$$t_{max} = 2^{M+1}/f_{clock} = 2^{11}/1 \text{ MHz} = 2048 \text{ }\mu s.$$

PROBLEM 7.17

KNOWN: D/A converter: M = 8
  E_o = 3.58 V if register = 10110011

FIND: E_o if register = 01100100.

ASSUMPTION: Straight binary code.

SOLUTION

10110011 = 179_{10} or 179 counts. For a known E_o this is equivalent to:

Q = 3.58 V/179 = 0.020 V

Then, for 01100100 = 100_{10} or 100 counts: E_o = (0.020 V)(100) = 2.000 V.

PROBLEM 7.18

KNOWN: Successive approximation converter: $M = 4$; $E_{FSR} = 10$ V; $E_i = 4.9$ V

FIND: E^*. Required M for $Q \leq 2.5$ mV.

SOLUTION

(i) Resolution: $Q = 10 \text{ V}/2^4 = 0.625$ V So with $E_i = 4.9$ V, the number of register counts is 4.9 V/0.625 V = 7.8. Referring to Figure 7.7, if we assume the A/D uses a straight encoding scheme whereby $e_Q = 1$ bit $= 0.625$ V, the register will show 7.8 → 7 counts, or $7_{10} = 0111$. Hence, $E^* = (7)(.625 \text{ V}) = 4.375$ V. Note that $E_i - E^* < e_Q$.

If the A/D used a bias shift encoding whereby $e_Q = \pm 1/2$ bit $= \pm 0.3125$ V, the register will show 7.8 → 8 counts, or $8_{10} = 1000$. Hence, $E^* = 5.000$ V. Again, $E_i - E^* < e_Q$.

(ii) For $|E_i - E^*| = e_Q < 2.5$ mV: requires $M = 12$ (straight encoding scheme) or $M = 11$ (bias shifted scheme).

PROBLEM 7.19

SOLUTION

Resolution: $Q = E_{FSR}/2^M$ regardless of the conversion method.

Maximum Conversion time: For a given clock speed, f_{clock},

Successive Approximation: $t = M/f_{clock}$

Ramp: $t = 2^M/f_{clock}$

Parallel: $t = 1/f_{clock}$

For any converter, resolution improves with bit number equally. The trade-off is seen in the required conversion times. Ramp times increase exponentially with bit number. Parallel method conversion times are essentially independent of bit number, but their component costs increase exponentially with bit number.

PROBLEM 7.20

KNOWN: A/D converter: $M = 10$; $E_{FSR} = 10$ V; register = 1010 1101 11

FIND: E_i

SOLUTION

For a straight binary code, 1010 1101 11 = 695_{10} which is equivalent to

$$E^* = (695)Q = (695)(10/2^{10}) = 6.787 \text{ V}$$

Assuming that $e_Q = Q = 1$ LSB,

$$E_i = E^* + 1 \text{ LSB} = 6.787 + 1 \text{ LSB} \text{ or } 6.787 \leq E_i \leq 6.796 \text{ V}$$

Assuming that $e_Q = \pm Q/2 = \pm 1/2$ LSB,

$$E_i = E^* \pm 1/2 \text{ LSB} \text{ or } 6.782 \leq E_i \leq 6.792 \text{ V}$$

PROBLEM 7.21

KNOWN: A/D converter: $M = 8$; $E_{FSR} = 10$ V; register = 1010 1011

FIND: E_i

SOLUTION

For two's complement code, 1010 1011 = -85_{10} {i.e. $-(128 - 43)$} which is equivalent to

$$E^* = (-85)Q = (-85)(10/2^8) = -3.320 \text{ V}$$

Assuming that $e_Q = Q = 1$ LSB,

$$E_i = E^* + 1 \text{ LSB} = -3.320 + 1 \text{ LSB} \quad \text{or} \quad -3.320 \geq E_i \geq -3.359 \text{ V}$$

Assuming that $e_Q = \pm Q/2 = \pm 1/2$ LSB,

$$E_i = E^* \pm 1/2 \text{ LSB} \quad \text{or} \quad -3.339 \leq E_i \leq -3.300 \text{ V}$$

PROBLEM 7.22

KNOWN: Dual slope A/D converter: $M = 12$; $f_{clock} = 10$ kHz;
register = 2011_{10}

FIND t_c

SOLUTION

The actual conversion time based on the register value of 2011 counts is

$$t_c = (2)(2011)/10 \text{ kHz} = 0.4022 \text{ s}$$

PROBLEM 7.23

KNOWN: Single ramp A/D converter: $M = 8$; $f_{clock} = 1$ MHz;
register = 173_{10}

FIND t_c

SOLUTION

The actual conversion time based on the register value of 173 counts is

$$t_c = (173)/1 \text{ MHz} = 173 \text{ } \mu s$$

PROBLEM 7.24

KNOWN: Successive approximation A/D converter: $M = 8$; $E_{FSR} = 10$ V
$E_i = 6.2$ V

FIND: Register value

SOLUTION

The resolution is: $Q = 10 \text{ V}/2^8 = 0.039$ V So each register bit value counts as 39 mV. For an input of 6.2V, the register count would be 6.2 V/0.039 V = 158.9. If we account for register threshold errors, we take 158.9 ➡ 159_{10} = 10011111 (straight binary).

The comparator will see $E^* = 6.162$ V after 7 steps and 6.201 V after 8 steps.

PROBLEM 7.25

KNOWN: Balance scale:
  input range: 0 to 5 kg
  output range: 0 to 3.50 mV
  Recorder:
  $M = 12$; $E_{FSR} = 10$ V

FIND: Appropriate amplifier gain, G.

SOLUTION

The balance scale has a sensitivity: $K = 3.5 \text{ mV}/5 \text{ kg} = 0.7 \text{ mV/kg}$

The recorder has a resolution: $Q = 10 \text{ V}/2^{12} = 2.44 \text{ mV}$

It is clear that the recorder resolution is large compared to the balance output. Amplification of the balance output prior to recording is in order. To improve the recorder resolution to within, say, 1% of the scale full scale output we will need, $Q = (0.01)(3.5 \text{ mV}) = 35 \ \mu V$. To achieve this,

$$G = 2.44 \text{ mV}/35 \mu V = 70$$

A minimum gain of about $G = 70$ is required. A higher gain will further improve resolution. Note that when an amplifier is used: $Q = E_{FSR}/(G)(2^M)$.

PROBLEM 7.26

KNOWN: $f \approx 2$ Hz
$N\delta t = 10$ s
Recorder: $M = 12$; $E_{FSR} = 10$ V

FIND: f_S, $y(t)$, δf, f_N

SOLUTION

(i) To meet the Sampling Theorem, $f_s > 2f_m$. Using $F_m = f = 2$Hz, this gives $f_s > 4$ Hz. But because f_m is not known exactly, we should set f_s away from 4 Hz. The minimum sample rate will also depend on N and the extent of spectral leakage (see amplitude ambiguity). Trial runs should be made to achieve acceptable results. For a 10 s block of data, i.e. $N\delta t = N/f_s = 10$ s, so that with we have $N = 10f_s$. As an example, let $f_s = 20$ Hz and $N = 200$.

(ii) $y(t) = C_1 \sin 2\pi ft = 2 \sin 4\pi t$ [V]

(iii) From (i): $\delta f = 1/N\delta t = f_s/N = 0.1$ Hz; $f_N = f_s/2 = 10$ Hz. A 10 Hz low pass filter should be used in front of the A/D converter.

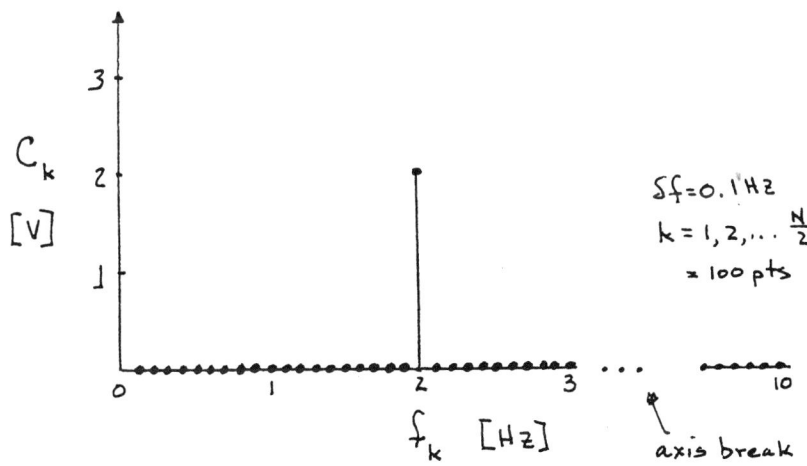

PROBLEM 7.27

KNOWN: DAS: f_s = 20 kHz (see Note below); M = 12; $N\delta t = N/f_s$ = 5 s
Computer: M = 16

FIND: Memory requirement

SOLUTION

Given that 5 s of data are acquired: N = (5s)(20,000/s) = 100,000 So 100,000 12-bit data points are acquired.

A 16-bit computer stores each integer number as a 16-bit word comprised of 2, 8-bit bytes. This requires that each 12-bit integer data point be stored as 2 bytes. So it takes 200,000 bytes to store the data. Now 1 kbyte = 1 kB = 1024 bytes. So 200,000 bytes requires 195.3 kB of memory.

Note: The first printing shows f_s = 20.000 Hz. At that rate, N = (5s)(20Hz) = 100 points, roughly 0.097 kB.

PROBLEM 7.28

KNOWN: DAS: $M = 12$; $f_s = 100$ MHz
Computer: 8 MB of 32-bit memory

FIND: $(N\delta t)_{max}$

SOLUTION

Direct Memory Access (DMA) writes data directly from acquisition into memory. If each 12-bit data point is stored as a 32-bit (4, 8-bit bytes) word, then the maximum storage is N = 8 MB = 8,192,000 bytes.

If $f_s = 100$ MHz, then $(N\delta t)_{max} = 0.08192$ s. If $f_s = 100$ kHz, then $(N\delta t)_{max} = 81.92$ s.

COMMENT

When multiple channels are sampled, the number of data points acquired per unit time is increased by the number of channels. The sample duration available is decreased by roughly the number of channels (the amount is not exact because some time is required to flip between channels). For example, a race car test might involve some 64 different sensors. Even at 100 Hz, that is 6400 data points per second.

PROBLEM 7.29

KNOWN: Square wave: $T_1 = 1$ s (Example 2.3)

FIND: N, f_s, f_c

SOLUTION

For the square wave of example 2.3, the first five non-zero terms show a frequency of $f_n = 2n\pi t/T_1$ for $n = 1, 3, 5, 7, 9$. So $f_m = 9$ Hz. The amplitudes show: $C_n = C_1/n$ for $n = 1, 3, ..., 9$. Hence: $f_s > 18$ Hz.

We want $mT_1 = N/f_s$ to minimize leakage. This gives: $N = mf_s$. Any combination of N and f_s will do provided $f_s > 18$ Hz.

For example: $f_s = 32$ Hz and $N = 32m$ where m is a positive number. If we let $m = 8$, $N = 256$. For this case, set $f_c = f_N = f_s/2 = 16$ Hz.

$$y(t) \approx \sum_{n=1}^{9,2} \frac{C_1}{n} \sin \frac{2n\pi}{T_1} t \quad \text{gives first five terms}$$

PROBLEM 7.30

KNOWN: Triangle wave: $T_1 = 2$ s

FIND: f_s, N, f_c

SOLUTION

The first seven non-zero terms have the form:

$$y(t) \approx (4C_1/\pi)\cos \pi t + (4C_1/3\pi)\cos 3\pi t + \ldots + (4C_1/13\pi)\cos 13\pi t$$

Hence, $f_m = 13/2 = 6.5$ Hz, so that $f_s > 13$ Hz.

We want $mT_1 = N/f_s$ to minimize leakage or $N = 2f_s m$.

Any combination will work provided $f_s > 13$ Hz. For example, $f_s = 16$ Hz, $N = 256$ and $f_c = f_s/2 = 8$ Hz.

PROBLEM 7.31

SOLUTION

Using examples from Problem 7.29: f_s = 32 Hz, N = 256. The discrete Fourier transform will return N/2 = 128 amplitudes corresponding to the N/2 frequencies spaced $\delta f = f_s/N$ = 0.125 Hz apart up to and including $f_N = f_s/2$ = 16 Hz. to generate a plot, we assume that C_1 = 1 (any value will do) and perform the DFT.

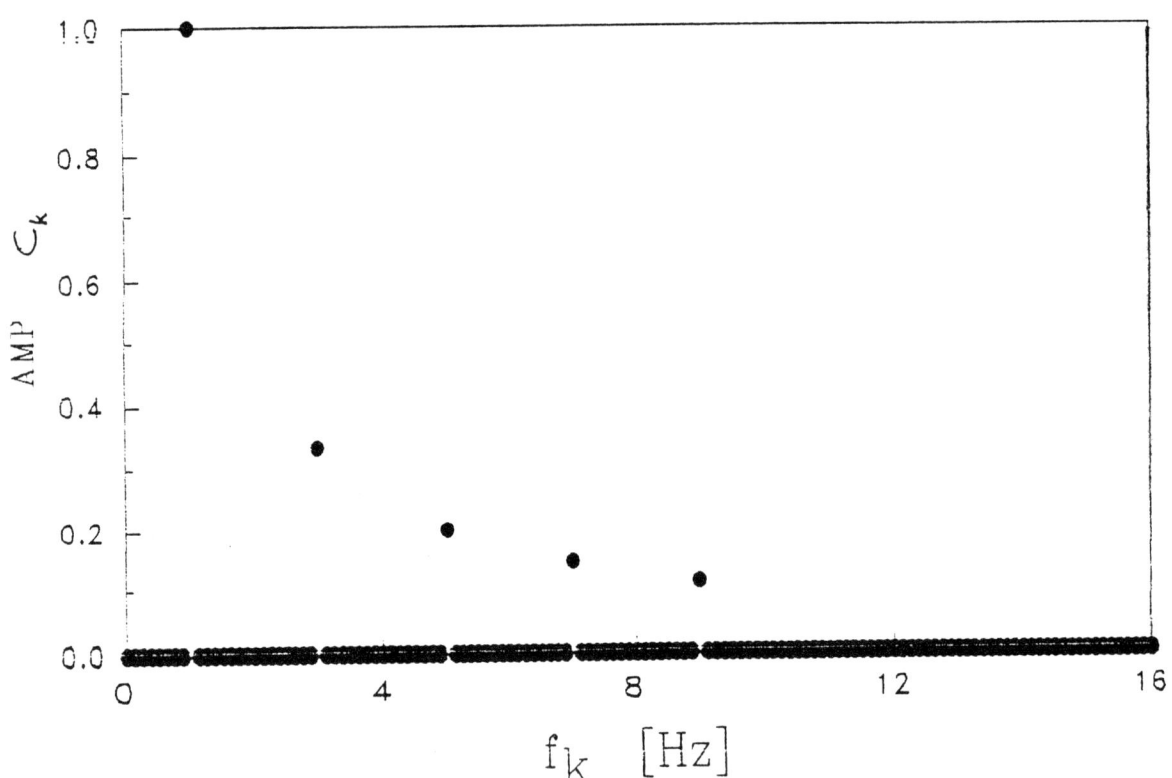

PROBLEM 7.32

SOLUTION

Using examples from Problem 7.30: $f_S = 16$ Hz, $N = 256$. The discrete Fourier transform will return $N/2 = 128$ amplitudes corresponding to the $N/2$ frequencies spaced $\delta f = f_S/N = 0.0625$ Hz apart up to and including $f_N = f_S/2 = 8$ Hz. To generate a plot, we assume that $C_1 = 1$ (any value will do) and perform the DFT.

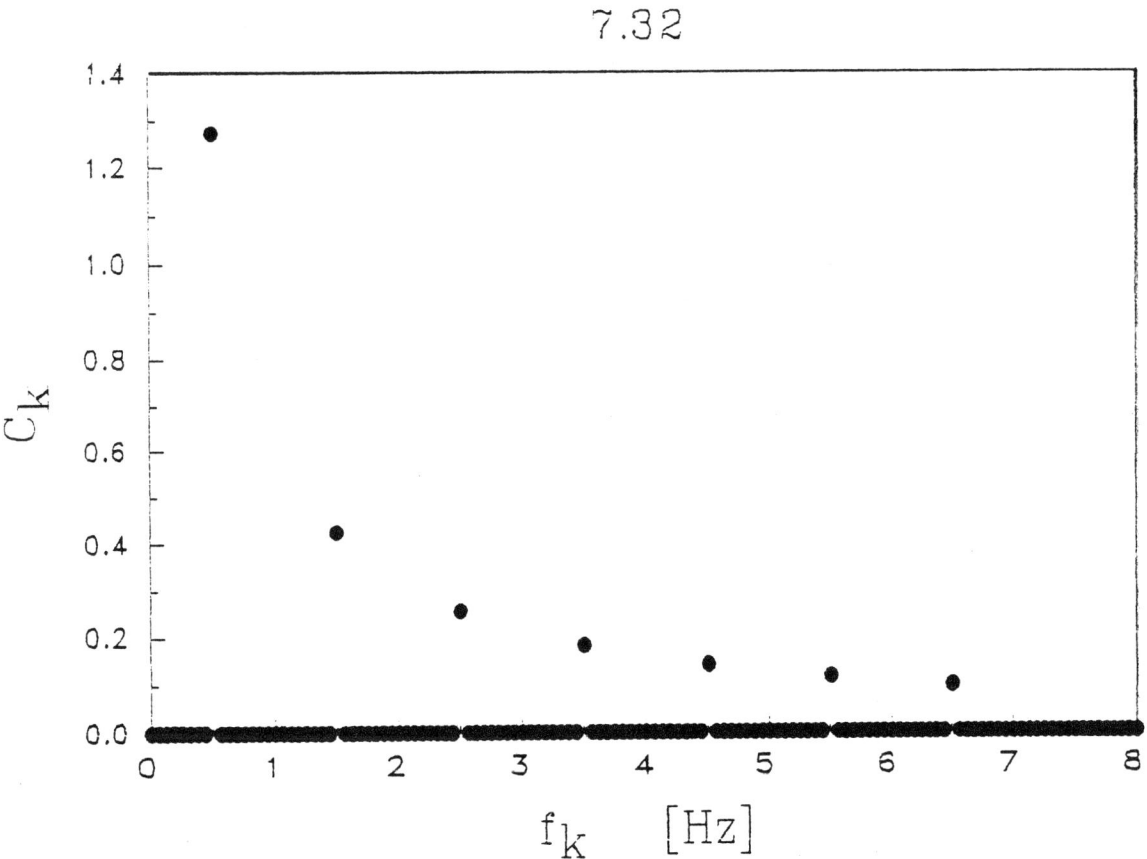

PROBLEM 7.33

KNOWN: RC filter
k = 1
f_c = 100 Hz

FIND: Attenuation at 10, 50, 75, and 200 Hz

ASSUMPTIONS: Filter is of the low pass, Butterworth type

SOLUTION

For a low pass RC Butterworth filter,

$$M(f) = 1/[1 + (f/f_c)^{2k}]^{0.5}$$

Recall that "attenuation" means reduction. In the context of a filter, a device which reduces (M(f) < 1) the output amplitude of targeted frequencies (M(f) < 1), it must correspond to a negative value of dynamic error. Dynamic error is given by

$$\delta(f) = M(f) - 1$$

Recall also that "gain" refers to a positive value of dynamic error.

For a low pass, RC Butterworth filter, M(f) ≤ 1 always, which is consistent with a 1st order system behavior.

f [Hz]	M(f)	δ(f)	attenuation [%]
10	0.995	-0.005	0.5
50	0.894	-0.105	10.5
75	0.800	-0.200	20.0
200	0.447	-0.553	55.3

PROBLEM 7.34

KNOWN: Low-pass LC Bessel filter with $k = 3$; $f_c = 100$ Hz

FIND: Attenuation at 10, 50, 75, 200 Hz

SOLUTION

A three-stage, low-pass Bessel filter will have the transfer function (6.62)

$$G(s) = a_0/[a_0 + a_1 s + a_2 s^2 + a_3 s^3] = a_0/D_3(s)$$

For unit cut-off frequency, $D_3(s)$ has the form of (6.60a), with $k = 3$

$$D_k(s) = D_3(s) = (2k - 1)D_2(s) + s^2 D_1(s)$$

with $D_1(s) = s + 1$ and $D_2(s) = s^2 + 3s + 3$. The three stage, unit cut-off frequency transfer function is

$$G(s) = 15/[s^3 + 6s^2 + 15s + 15]$$

With $\omega_c = 2\pi f_c = 628$ rad/s, we substitute s/ω_c for s and solve for $s = j\omega$ to obtain

$$G(j\omega) = 15\omega_c^3/[-j\omega^3 - 6\omega^2\omega_c + 15j\omega\omega_c^2 + 15\omega_c^3]$$

$$= M(\omega)e^{j\Phi(\omega)}$$

From which, the magnitude

$$M(\omega) = 15\omega_c^3/[(15\omega\omega_c^2 - \omega_c^3)^2 + (15\omega_c^3 - 6\omega^2\omega_c)^2]^{0.5}$$

f	M(f)	Attenuation
10	1	0%
50	0.98	2%
75	0.94	6%
100	0.63	37%

PROBLEM 7.35

KNOWN: RC low-pass Butterworth filter
-3 dB $\leq$ M($0 \leq f \leq 5$ kHz) and M($f \geq 10$ kHz) < -30 dB

FIND: Values for R, C and k

SOLUTION

This problem is an open-ended design. One possible solution follows.
For a low-pass, Butterworth filter,

$$M(f) = 1/[1 + (f/f_c)^{2k}]^{0.5}$$

where $f_c = 1/2\pi\tau$ and $\tau = RC$. We can set $f_c = 5$ kHz, such that M(5kHz) = 0.707 = -3 dB, and meet one constraint of the design. This gives,

$$\tau = RC = 1/2\pi(5000 \text{ Hz}) = 31.8 \text{ } \mu s$$

One design combination could use R = 3,180 Ω and C = 10 pF.

The next step in the design is to determine the number of filter stages required to meet the attenuation constraint at 10 kHz. For at least 30 dB attenuation

$$-30 \text{ dB} \geq 20 \log M(10000 \text{ Hz})$$

then,

$$M(10000 \text{ Hz}) \leq 0.0316 = 1/[1 + (10 \text{ kHz}/5 \text{ kHz})^{2k}]^{0.5}$$

By trial and error, k $\geq$ 5.

k	M(10000 Hz)
1	0.45
3	0.12
5	0.0312

PROBLEM 7.36

KNOWN: y(t) is a complex periodic waveform passed through a low pass RC filter and then sampled discretely.
$f_s = 500$ Hz
Filter cut-off: $f_c = 250$ Hz

FIND: Maximum frequency component, f_m, of y(t) that can be sampled with $\delta(f) \leq 0.10$

ASSUMPTIONS: One stage RC Butterworth filter

SOLUTION

The input signal has the form

$$y(t) = A_o + \sum_n C_n \sin(2\pi n f_1 t + \Phi_n)$$

We seek the value for $f_m = n f_1$ such that $\delta(f_m) = M(f_m) - 1 \leq 0.10$.

$$M(f) = 1/[1 + (f/f_c)^{2k}]^{0.5}$$

Setting $f_c = 250$ Hz, $M(f_m) = 0.9$ and assuming $k = 1$, then

$$0.9 \leq 1/[1 + (f_m/250)^2]^{0.5}$$

and $f_m \leq 121$ Hz.

COMMENT

This solution does not consider the amplitude error involved in sampling y(t) at 500 Hz. We must assume by the problem statement that this sample rate and the sample period have been properly chosen to provide a minimum error in reconstruction of the spectral amplitudes and frequencies.

PROBLEM 7.37

KNOWN: A 500 Hz component is to be removed from an analog signal and passed through an A/D converter.
$M = 8$; $E_{FSR} = 10$ V; $f_s = 200$ Hz

FIND: Low-pass anti-alias filter with $M(500 \text{ Hz}) \leq e_Q$.

SOLUTION

Quantization error: $e_Q = \pm 0.5[E_{fsr}/2^M] = \pm 0.5[10\text{V}/256] = \pm 0.01953$ V

With $f_s = 200$ Hz, an appropriate anti-alias filter will have $f_c \leq 100$ Hz. We set $f_c = 100$ Hz. Then, for the filter

$$\tau = 1/2\pi f_c = 1/2\pi(100 \text{ Hz}) = 1.59 \text{ ms}$$

One combination has $R = 1590 \, \Omega$ and $C = 1 \, \mu\text{F}$.
To select the number of filter stages,

$$M(500) = 1/[1 + (f/f_c)^{2k}]^{0.5} = 1/[1 + (500 \text{ Hz}/100 \text{ Hz})^{2k}]^{0.5}$$

Recall that the magnitude ratio is defined by the ratio of the output amplitude to the input amplitude. The quantization error is fixed at 0.01953 V. The output amplitude at 500 Hz depends on the input amplitude and M(500). Solving for k in terms of an input amplitude, A:

$$k = [2\log A + 5.597]/1.3979$$

A [V]	k
0.02	2
0.10	3
1.00	4
10.00	6

So if $A = 1$ V, a two-stage filter will work; if $A = 10$ V, six stages are needed.

PROBLEM 7.38

KNOWN: Sensor-transducer: Thermocouple
Thermocouple measures $50 \leq T \leq 70\ °C$ and outputs $2.585 \leq E \leq 3.649$ mV.
Signal is digitized using an A/D converter:
$M = 12$; $E_{fsr} = 10$ V (ie -5 to +5 V, bipolar); SNR = 40 dB

FIND: (a) e_Q/E
(b) gain (G) required to reduce a to 5% or less
(c) estimate SNR in (b)

SOLUTION

(a) The quantification error of an M-bit device is

$$e_Q = \pm 0.5[E_{fsr}/2^M] = \pm 0.5[10\ V/4096] = \pm 1.22\ mV$$

One can expect the relative quantification error to vary from:

$$e_Q/E = 1.22\ mV/2.585\ mV = 0.472$$

or 47% at 50°C, to

$$e_Q/E = 1.22\ mV/3.649\ mV = 0.33$$

or 33% at 70°C. Both values are significantly large.

(b) One means to reduce the relative quantization error is through amplification of the analog signal prior to quantization. To achieve 5% or less error requires an input signal of the magnitude,

$$E = e_Q/0.05 = 1.22\ mV/0.05 = 24.40\ mV$$

At 50°C (the smallest voltage quantized), this requires a linear amplifier gain of

$$G = E_o/E_i = 24.40\ mV/2.585\ mV = 9.44 \approx 10$$

Or roughly, an amplifier having a linear gain of 10.

(c) Any signal is composed of a magnitude attributable to deterministic signal, E_s and a magnitude attributable to noise, E_n.

$$\text{SNR} = 20\log E_s/E_n$$

With SNR = 40 dB, $E_s = 100 E_n$.

COMMENT

The signal level is 100 times greater than noise prior to amplification. During amplification, all information within the frequency range of the amplifier will be scaled by the gain of the amplifier. So the SNR will not change. However, quantization error will manifest itself as random noise. We see in part (a) that the signal will be masked by this large noise level. But this noise is added only after quantization. Amplification of the analog signal occurs before quantization. We see in part (b) that the signal level is raised to a level where it will not be masked by the quantization noise. So a benefit in amplification is a reduction in the relative quantization error.

PROBLEM 7.39

KNOWN: A/D Converter: $M = 8$ or 12; $E_{fsr} = 10$ V; $0 < f_s \leq 100$ Hz

FIND: Specify M, f_s, amplifier gain G, and filter (type, f_c and k). Estimate e_Q and $N\delta t$ expected for your choice.

SOLUTION

This design problem has an open-ended solution path. To demonstrate, one possible solution is presented for each case.

(a) $E(t) = 2\sin 20\pi t$ V

The input signal has an amplitude of $A = 2$ V and a single frequency of $f_1 = 10$ Hz. Accordingly, we will want

(1) $f_s > 2f_m$ that is $f_s > 20$ Hz
(2) $mf_1 = N\delta t$ $m = 1, 2, ...$

to correctly reconstruct both signal frequency and amplitude in the time discrete series. If we set $f_s = 40$ Hz so as to satisfy (1), we should sample the signal at data rates in multiples of 4, i.e. from (2) $N = 4m$, $m = 1, 2, ...$. The sample period, $N\delta t$, will then be consistent with (1) and (2).

The 12-bit A/D converter is chosen for its better resolution and quantization error:

$$Q = E_{fsr}/2^M = 10 \text{ V}/4096 = 2.44 \text{ mV}$$

$$e_Q/E = \pm 0.5Q/A = \pm 0.00122 \text{ V}/2 \text{ V} = \pm 0.0006$$

Because $A = 2$ V, which fits within the ± 5 V range of the converter, AND because the relative quantization error is so small, no amplification is required.

An anti-alias filter is always required. A low-pass, RC Butterworth filter is selected for its flat bandpass characteristics. Setting $f_c = f_s/2 = 20$ Hz,

$$\tau = RC = 1/2\pi f_c = 1/2\pi(20 \text{ Hz}) = 7.96 \text{ ms}$$

Values of $R = 7960\ \Omega$ and $C = 1\mu F$ are chosen to minimize signal loading. The effect of stage number on signal attenuation at $f = 10$ Hz is:

$$\delta(f) = M(f) - 1 = 1/[1 + (f/f_c)^{2k}]^{0.5} - 1$$

k	M(10 Hz)	δ(10 Hz) [%]
1	0.895	-10.5
3	0.992	- 0.8

Set k = 3 to provide an attenuation due to the filter of less than 1% at 10 Hz.

(b) $E(t) = 1.5\sin \pi t + 20\sin 32\pi t - 3\sin(60\pi t + \pi/4)$ V

The input signal contains amplitudes of A_1, A_2 and A_3 with frequencies of $f_1 = 0.5$, $f_2 = 16$ and $f_3 = 30$ Hz, respectively. The maximum frequency in the signal, f_m, is 30 Hz. The signal does not contain a single fundamental frequency. We need

(1) $f_s > 2f_m$ or $f > 60$ Hz

(2) $mT = N\delta t$ $m = 1, 2, ...$

Criterion (2) can be rewritten as $m/f = N/f_s$. In order to meet both criteria for this signal requires a minimum sample rate of $f_s = 240$ Hz with N = 480 (i.e. Nf/f_s = 1, 32 and 60, an exact integer multiple of each of the three frequencies, respectively). Alas, the A/D converter does not have such a capability! As one compromise, setting $f_s = 80$ Hz with N = 160m will meet the criteria for both the f_1 and f_2 components, but not for f_3. We should expect leakage about f_3 in the discrete time series representation. An alternative approach is to use trial and error on f_s and N until leakage is reduced to acceptable levels. components.

The 12-bit A/D converter is chosen for its better resolution.

$$Q = E_{fsr}/2^M = 10 \text{ V}/4096 = 2.44 \text{ mV}$$

Amplitude A_2 will saturate the A/D converter. We choose a linear amplifier with a gain of G = 0.2 to keep the signal well within range. The relative quantization error becomes:

$$e_Q/E = \pm 0.5Q/GA = \pm 0.00122 \text{ V/GA}$$

for values of 0.004, 0.0003, and 0.002 for the respective frequencies. Because this is below 1%, a value which we judge sufficient, The amplifier gain seems appropriate.

An anti-alias filter is always required. A low-pass, RC Butterworth filter is selected for its flat bandpass characteristics. Setting $f_c = f_s/2 = 40$ Hz,

$$\tau = RC = 1/2\pi f_c = 1/2\pi(40 \text{ Hz}) = 3.98 \text{ ms}$$

Values of R = 3980 Ω and C = 1μF are chosen to minimize signal loading. The effect of stage number on signal attenuation is:

$$\delta(f) = M(f) - 1 = 1/[1 + (f/f_c)^{2k}]^{0.5} - 1$$

k	M(0.5)	M(16)	M(30)	$\delta(30)$ [%]
1	1	0.93	0.80	-20
3	1	1	0.92	-8
5	1	1	0.97	-3
7	1	1	0.99	-0.9

Set k = 7 to keep attenuation below 1% at all frequencies (see COMMENT).

COMMENT

This is a large number of stages for a passive filter circuit but it will work. Note that this k problem enters from the limitation on sample rate. If f_s, and then f_c, could be increased, the number of stages could be reduced. The disadvantage in using sample rates considerably higher than $2f_m$ is that noise in the signal above f_m is also sampled.

(c) P(t) = -10sin 4πt + 5sin 8πt psi; K = 0.4 V/psi

The voltage signal sensed by the data acquisition system will be

E(t) = KP(t) = -4sin 4πt + 2sin 8πt V

The input signal has amplitudes A_1 = 4 V and A_2 = 2 V with f_1 = 2 Hz and $f_2 = 2f_1$, respectively. The maximum frequency is f_m = 4 Hz. We want

(1) $f_s > 2f_m$ or $f_s > 8$ Hz
(2) $mT_1 = N\delta t$ m = 1, 2, ...

Criterion (2) can be rewritten as $m/f_1 = N/f_s$. If f_s = 10 Hz, then we must sample at data rates of N = 2m.

The 12-bit A/D converter is chosen for its better resolution and small relative quantization error:

$$Q = E_{fsr}/2^M = 10 \text{ V}/4096 = 2.44 \text{ mV}$$
$$e_Q/E = \pm 0.5Q/A = \pm 0.00122 \text{ V/A}$$

For A_1 = 4 V and A_2 = 2 V, e_Q/E = ± 0.0003 and 0.006, respectively. Because A_1 and A_2 fit within the ±5 V range of the converter AND because the relative quantization error is so small, no amplification is required.

An anti-alias filter is always required. A low-pass, RC Butterworth filter is selected for its flat bandpass characteristics. Setting $f_c = f_s/2 = 5$ Hz,

$$\tau = RC = 1/2\pi f_c = 1/2\pi(5 \text{ Hz}) = 31.8 \text{ ms}$$

Values of R = 3180 Ω and C = 10μF are chosen to minimize signal loading. The effect of stage number on signal attenuation is:

$$\delta(f) = M(f) - 1 = 1/[1 + (f/f_c)^{2k}]^{0.5} - 1$$

k	M(2)	M(4)	δ(4) [%]
1	0.98	0.78	- 22
5	1	0.95	- 5
7	1	0.98	- 2

We set k = 5 to keep attenuation below 5%.

COMMENT

One way to improve on this attenuation number without increasing the number of required filter stages is to choose a higher value for f_s. This enables selecting a higher f_c. For example, with $f_s = 50$ Hz, N = 25m, we could set $f_c = 25$ Hz requiring k = 1 to reduce attenuation well below 1%. However, the disadvantage in using sample rates considerably higher than $2f_m$ is that noise in the signal above f_m is also sampled. This may or may not be a problem depending on the application.

PROBLEM 7.40

KNOWN: DAS: 4 channels; M = 8; f_s = 4000 Hz; $N\delta t$ = 100s.

FIND: N, effective sample rate for channel one.

SOLUTION

Here we sample 4 channels at 4000 Hz. The effective sample rate for channel one will be about 4000/4 = 1000 Hz. If the sampling continues for 100s then, N = (4000/s)(100s) = 400,000 (8-bit) data points. This will require 400,000 bytes or (at 1024 = 1 kB) 390.6 kB of memory.

PROBLEM 7.41

KNOWN: Transducers: ± 1 V output; ± 25 cm H_2O input
DAS: M = 10; E_{FSR} = 10 V; 4 MB memory; 10 min. battery life.

FIND: N, f_s, f_c, G

SOLUTION

The transducer sensitivity is: K = 2V/50 cm H_2O = 0.04 V/cm H_2O.
The A/D resolution is: Q = $10V/2^{10}$ = 0.00976 V. This can be expressed as Q = 0.00976 V/0.04 V/cm H_2O = 0.244 cm H_2O.
The sensitivity meets problem constraints. However, an analog amplifier between the transducer and the A/D with a gain of G = 5 will take full advantage of the A/D range and improve resolution:
Q = $10V/(5)(2^{10})$ = 0.00195 V = 0.0488 cm H_2O.
Underhood aerodynamic pressures are affected by car velocity and engine operation. Assuming a constant car speed, underhood pressures will fluctuate about a constant mean pressure. A high RPM 8-cylinder engine ($\approx$ 8000 RPM) would produce frequencies of about 16 Hz or so. Using 20 Hz as a maximum, we need f_s > 40 Hz. Further, we want $mT_1 = N/f_s$. Setting f_s = 50 Hz, we can set N = 30,000 data points over 10 minute total sample period. We should use an anti-alias filter set at $f_c = f_s/2$ = 25 Hz.

COMMENT

Unless we are interested in engine intake pressures, the fluctuations will be very small in amplitude throughout the underhood region and a much lower sample rate can be used with an appropriate anti-alias filter, such as f_s = 10 to 20 Hz.

PROBLEM 8.1

KNOWN/FIND: Define and discuss the significance of:
  a) temperature scale
  b) temperature standards
  c) fixed points
  d) interpolation

SOLUTION:

a) temperature scale - an established relationship for assigning numerical values to measures of temperature. The absolute temperature scales are:

  the Rankine scale, for U.S. customary units
  the Kelvin scale, for SI units

b) temperature standards - a formally adopted and recognized means for practical realization of temperature measurement. Standards provide a means for the measurement of temperature which can be reproduced and agrees with the thermodynamic definition of temperature.

c) fixed points - identifiable and experimentally reproducible conditions which are associated with a certain temperature (numerical value). See Table 8.1.

d) interpolation - a method for determining temperatures other than those defined by fixed points on a temperature scale. For the majority of applications, the interpolating instrument is a platinum RTD.

PROBLEM 8.2

KNOWN: An apparatus to produce phase equilibrium points is required.

FIND: Describe the conditions necessary to establish phase equilibrium points. Identify the effects of elevation, weather and material purity.

SOLUTION:

Other than the vapor-pressure-temperature points for helium and hydrogen, the fixed points for ITS-90 are freezing points, melting points or triple points.

triple point - The procedure for calibrating a thermometer at a triple point is:
1. completely freeze the sample (an appropriate mass of material) in a closed container
2. experimentally determine the energy required to melt the sample
3. re-freeze the sample
4. Add energy and record the thermometer output at 10, 20 40, 60, 70 and 80% melted. These readings should agree.

This procedure demonstrates that a container capable of preventing contamination of a sample of material, while allowing the removal and addition of energy, is required to establish the triple point for a material. Similar requirements are needed for melting/freezing points, with the notable exception that containers must generally be flexible, to accommodate thermal expansion. Representative values of measured temperatures agree to 0.1 mK.

A sample which is 99.9999% pure will produce measured temperatures over a phase change within 0.1 mK.

Weather and elevation should be eliminated through appropriate design of the experimental apparatus.

PROBLEM 8.3

KNOWN: A length of platinum wire having:
  length, l = 2 m
  diameter, D = 0.1 cm
  resistivity, $\rho_e = 9.83 \times 10^{-6}$ Ω-cm

FIND: The resistance of the wire, R

SOLUTION:

Since
$$R = \frac{\rho_e l}{A_c}$$
and
$$A_c = \frac{\pi}{4} D^2 = \frac{\pi}{4}(0.1)^2 = 0.0079 \text{ cm}^2$$
The resistance is found as

$$R = \frac{(9.83 \times 10^{-6})(2 \text{ m})(10^2 \text{ cm/m})}{0.0079 \text{ cm}^2} = 0.25 \text{ Ω}$$

COMMENT: An RTD would normally have a reasonably large resistance, on the order of 25 Ω. As such, a very small diameter wire or long length must be employed.

PROBLEM 8.4

KNOWN: A Wheatstone bridge and RTD as shown in Figure 8.35, with

$$\alpha = 0.003925°C^{-1}$$
$$R_o = 25 \, \Omega \text{ at } 0°C$$
$$R_1 = 41.485 \, \Omega \text{ for balanced conditions}$$

FIND: a) The temperature of the RTD

b) Compare the static sensitivity of this circuit to the circuit in Example 8.2

SOLUTION:

At balanced conditions

$$\frac{R_2}{R_3} = \frac{R_1}{R_4}$$

and when $R_1 = 41.485 \, \Omega$

$$= \frac{41.485}{R_4} \Rightarrow R_4 = 41.485 \, \Omega$$

From

$$R = R_o\left[1 + \alpha(T - T_o)\right]$$
$$41.485 = 25[1 + 0.003925(T - 0)]$$

we find

$$T = \frac{\frac{41.485}{25} - 1}{0.003925} \quad \text{and } T = 168°C$$

b) The static sensitivities are the same, since $R = R_1 (R_3/R_2)$ and $R_2 = R_3$ in both cases.

PROBLEM 8.5

KNOWN: A thermistor has a resistance of 20,000 Ω at 100°C.

$\beta = 3650°C$
$R_o = 20,000 \ \Omega$
$R = 500 \ \Omega$

FIND: The temperature corresponding to a thermistor resistance of 500 Ω.

SOLUTION:

From (8.11)

$$R = R_o e^{\beta\left(\frac{1}{T}-\frac{1}{T_o}\right)}$$

Letting $R_o = 20,000 \ \Omega$

$$R = 500 = 20,000 e^{3650\left(\frac{1}{T}-\frac{1}{373}\right)}$$

and

$$\ln 500 = \ln 20,000 + 3650\left(\frac{1}{T} - \frac{1}{373}\right)$$

Solving for T

$$T = 598.7 \ K = 325.7°C$$

PROBLEM 8.6

KNOWN: The uncertainty in temperature $u_T = \pm 0.005°C$.

FIND: The temperature corresponding to a thermistor resistance of 500 Ω.

ASSUMPTIONS: Initially assume that we wish to find the required uncertainty in resistance measurement as if it were the only contributor to the total uncertainty. In addition, this problem is open-ended to some extent, in that some nominal value of R_o must be assumed, or a range of values for R_o examined.

SOLUTION:

With

$$R = R_o[1 + \alpha(T - T_o)]$$

and $\alpha = 0.003925°C^{-1}$, we can express

$$u_T = \frac{\partial T}{\partial R} u_R$$

Then

$$T = \frac{(R/R_o - 1)}{\alpha} + T_o$$

$$\frac{\partial T}{\partial R} = \frac{1}{\alpha R_o}$$

Taking $R_o = 100$ Ω

$$\frac{\partial T}{\partial R} = \frac{1}{(0.003925)(100)} = 2.55°C/\Omega$$

and the uncertainty in resistance is

$$u_R = \pm 0.00196 \, \Omega$$

COMMENT: A parametric examination of the effect of the value of R_o on the uncertainty would contribute to the fundamental understanding of the measurement (see plot below). This is crucial at the design stage for a measurement system, and would provide information concerning the sensitivity of the design to R_o.

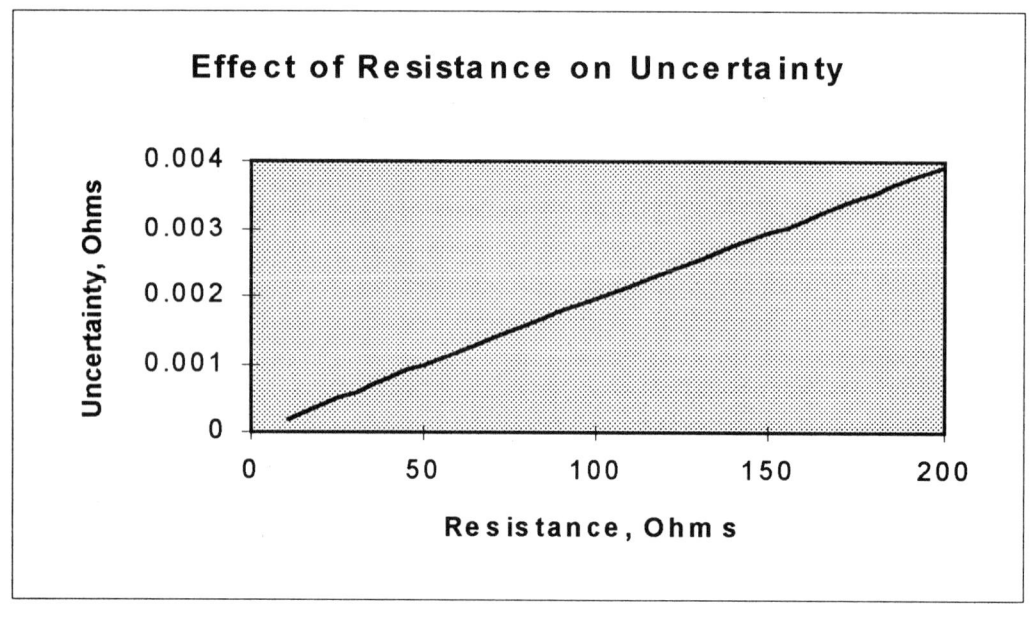

PROBLEM 8.7

KNOWN/FIND: Define and discuss the following terms related to thermocouple circuits:

a) thermocouple junction
b) thermocouple laws
c) reference junction
d) Peltier effect
e) Seebeck coefficient

SOLUTION:

a) thermocouple junction - electrical connection between two dissimilar metals which form a thermoelectric circuit.

b) thermocouple laws - observed behavior of thermoelectric circuits which allow the measurement of temperature using thermocouple circuits.

c) reference junction - an emf is present in a thermoelectric circuit having two junctions maintained at different temperatures. In order to measure temperature, one of the junctions must have a known temperature, and is called the reference junction.

d) Peltier effect - this phenomenon results from the conversion of electrical to thermal energy at a junction.

e) Seebeck coefficient - defines the relationship between temperature and emf for a thermocouple circuit.

PROBLEM 8.8

KNOWN: Measured emf at the potentiometer: emf = 10.668 mV

FIND: Measuring junction temperature

ASSUMPTIONS: The J-type thermocouple is within NIST standards and Table 8.6 may be utilized.

SOLUTION:

From Table 8.6 with an emf = 10.668 mV

the temperature is found as 198°C.

PROBLEM 8.9

KNOWN:
  a) Thermocouple circuit of Fig. 8.37a yields an emf of 7.947 with $T_{ref} = 0°C$

  b) $T_{ref} = 25°C$

  c) $T_{ref} = 0°C$ with copper extension wires installed

FIND: The indicated temperature

ASSUMPTIONS: NIST standard thermocouple behavior

SOLUTION:
a) from Table 8.6, T = 148.9°C
b) Knowing $emf_1 + emf_2 = emf_3$

$$7.947 \text{ mV} = emf_{0-25} + emf_{25-148.9}$$

with $emf_{0-25} = 1.277$ mV yields 6.67 mV for the output.

c) 148.9°C

PROBLEM 8.10

KNOWN: A J-type thermocouple referenced to 70°F. output emf = 2.878 mV with T_{ref} = 70°F.

FIND: The temperature of the measuring junction

ASSUMPTIONS: NIST Standard Behavior

SOLUTION:

To utilize Table 8.6, convert °F to °C

$$70°F = 21.1°C$$

and employing the Law of Intermediate Temperatures

$$emf_{0-21.1} = 1.076 \text{ mV}$$

$$emf_{0-T} = 1.076 + 2.878 = 3.954 \text{ mV}$$

and

$$T = 75.7°C$$

PROBLEM 8.11

KNOWN: A J-type thermocouple referenced to 0°C; output emf = 4.115 mV

FIND: The temperature of the measuring junction

ASSUMPTIONS: NIST Standard Behavior

SOLUTION: From Table 8.6 at an emf of 4.115 mV

$$T = 78.7°C$$

PROBLEM 8.12

KNOWN: An uncertainty level $u_T = 2°C$ at $200°C$ is required for a temperature measurement using a T-type thermocouple. The readout device used for this temperature measurement has:

accuracy: ± 0.5 C (e_1)
resolution: ± 0.1 C (e_2)

FIND: Determine if the uncertainty constraint is met.

ASSUMPTIONS: NIST Behavior

SOLUTION:

The elemental errors assoicated with the indicator (output stage) may be combined as

$$\sqrt{e_1^2 + e_2^2} = \sqrt{0.5^2 + 0.05^2} = \pm 0.503°C$$

This is the uncertainty which would result if the thermocouple exactly followed NIST Standard Behavior. The uncertainty due to variations from the NIST Standard are found from Table 8.5 as $\pm 1.0°C$ or $\pm 0.75\%$, whichever is larger. This yields $\pm 1.5°C$, and

$$u_T = \sqrt{1.5^2 + 0.503^2} = \pm 1.58°C$$

Yes, the uncertainty constraint is met.

PROBLEM 8.13

KNOWN: Thermocouple arrangement shown in Figure 8.21 with

> N = 3
> J-type thermocouples
> all junctions sense 3°C temperature difference
> Maximum variation from NIST - 0.8%
> Voltage measurement uncertainty ± 0.0005 V

FIND:

a) thermopile output for an average junction temperature of 80°C
b) the design stage uncertainty in measured temperature

SOLUTION:

The thermopile output will be 3 times that for 1 thermocouple sensing $\Delta T = 3°C$ at 80°C (approximated from Table 8.6)

$$\frac{\partial emf}{\partial T} = 0.053 \text{ mv-}°C^{-1}$$

The output is then

$$3 \times (3°C) \times (0.053 \text{ mV/°C}) = 0.477 \text{ mV}$$

b) First find the uncertainty in temperature which results from the uncertainty in the voltage measurement

$$u_T = \left(\frac{\partial T}{\partial emf}\right) u_{emf}$$

But $\partial T / \partial emf$ is the slope of the emf vs T curve at 80°C for the thermopile. For a single thermocouple, this slope is 1/0.053. For the thermopile, this slope is 1/[3(0.053)] or 0.053 mV/°C. This yields

$$u_T = \left(\frac{1°C}{0.053 \text{ mV}}\right)(0.05 \text{ mV}) = \pm 3.1°C$$

and there is a contribution due to the variation of the thermocouple from NIST standards, which is related to the uncertainty in the slope of the curve shown below

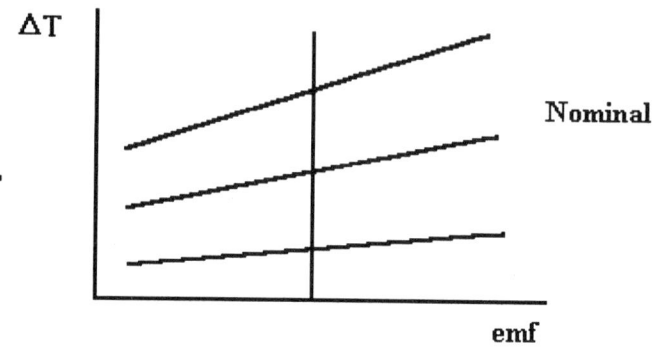

At $\Delta T = 3°C$, the uncertainty in temperature based on the $\pm 0.8\%$ yields ± 0.024 C, in the ΔT. Thus the resulting uncertainty is given by

$$u_T = \sqrt{0.024^2 + 3.1^2} = \pm 3.1°C$$

COMMENT:

The value of $\partial emf/\partial T$ is relatively insensitive to temperature; for example, at 400°C the value is 0.055 mV/°C. The uncertainty in voltage of ± 0.5 mV is unacceptable for most temperature measurements, since the resulting uncertainty is higher than the measured ΔT. However, if the number of junctions increased to 10, the resulting uncertainty would be $\pm 0.94°C$, which may be acceptable in many cases.

PROBLEM 8.14

KNOWN: Values of temperature and emf for a given reference temperature

FIND: Complete the table of values

ASSUMPTIONS: NIST Standard Behavior

SOLUTION:

Temperature [°C]

Measured	Reference	emf [mV]
100	0	5.269
-10	0	-0.501
100	50	2.684
96.6	50	2.5

PROBLEM 8.15

KNOWN: A thermopile having

4 junctions (N = 4)
$T_{ref} = 0°C$
$T = 125°C$
$u_{emf} = \pm 0.0001\ V = \pm 0.1\ mV$

FIND:

a) emf
b) N for an uncertainty of ±0.1 C

ASSUMPTIONS: NIST Standard Behavior
J-type thermocouple

SOLUTION:

a) for a single thermocouple

$$emf_1 = 6.634\ mV$$

Thus for the thermopile the output would be

$$4(6.634\ mV) = 26.536\ mV$$

b) The static sensitivity of the thermocouple at 125°C is approximately 0.055 mV/°C and

$$u_T = \left(\frac{\partial T}{\partial emf}\right) u_{emf}$$

Thus

$$0.1°C = \frac{1}{N(0.055)}\ °C/mV\left(0.0001 \times 10^3\ mV\right)$$

$$N = 18.2\ \text{or}\ 19$$

PROBLEM 8.16

KNOWN: A bimetallic thermometer serves as the sensing element in a thermostat for a residential heating/cooling system.

FIND:
Considerations for
- a) location for the installation of the thermostat
- b) effect of the thermal capacitance of the thermostat
- c) thermostats are often set 5°C higher in the air conditioning season

ASSUMPTIONS: Goal is to measure the air temperature inside the house

SOLUTION: Answers should address the following
1. Location should be on an inside wall to minimize conduction errors
2. Location should not be exposed to direct solar radiation, to prevent radiation errors
3. Location should not be in the direct flow from the HVAC system
4. The thermal capacitance of the bimetallic thermometer typically yields a time constant much shorter than required to regulate room temperature
5. Thermostats are typically set 5°C higher in summer primarily to save energy, but also to accommodate seasonal lifestyle changes.

PROBLEM 8.17

KNOWN: A J-type thermocouple is to be used at temperatures between 0 and 100°C. A single calibration point is available, at the steam point. Barometric pressure is 30.1 in. Hg, and the measured emf = 5.310 mV

FIND: Develop a calibration curve for this thermocouple

ASSUMPTIONS: (emf_{ref} - emf_{meas}) is linear from 0°C to 100°C

SOLUTION:

First, determine the steam point temperature for this barometric pressure

$$T_{st} = 212 + 50.422\left(\frac{30.1}{29.921} - 1\right) - 20.95\left(\frac{30.1}{29.921} - 1\right)^2$$

which yields T_{st} = 212.30°F = 100.17°C

A calibration curve can be plotted with the dependent variable as (emf_{ref} - emf_{meas}) where

emf_{ref} = NIST standard emf value (mV)

emf_{meas} = measured output from thermocouple (mV)

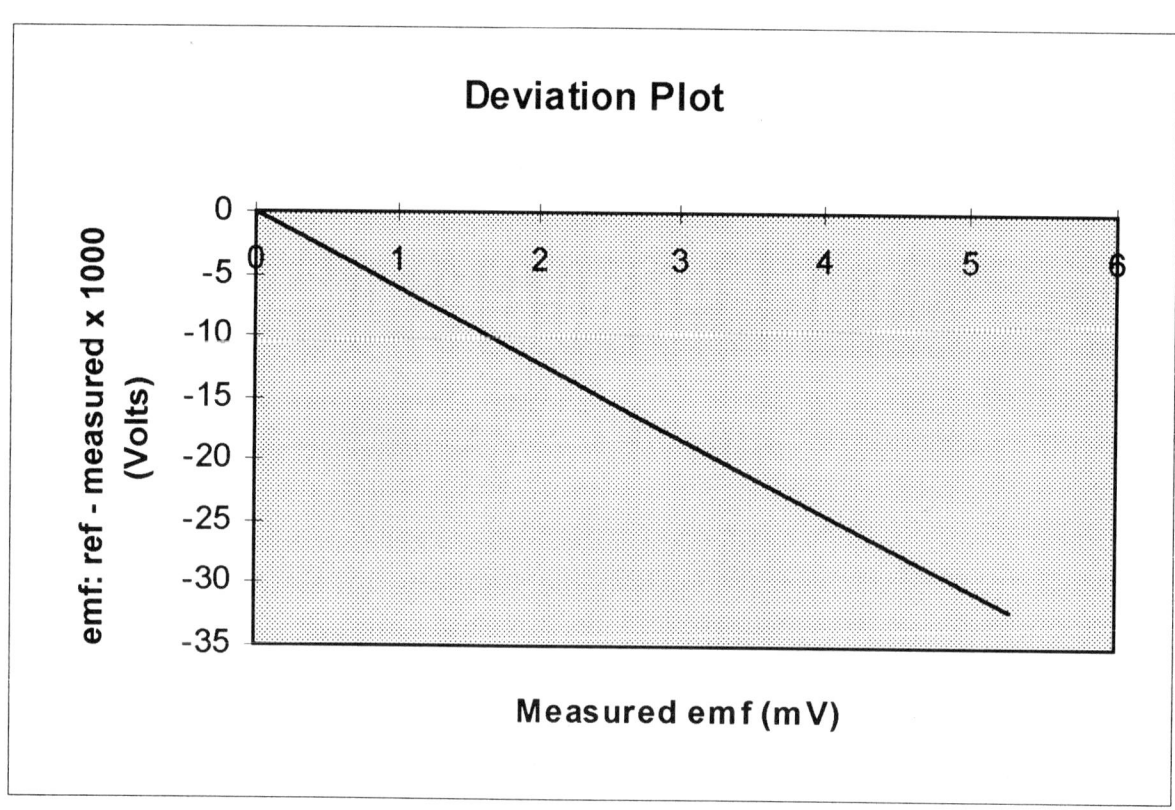

In this case, from Table 8.6

$$\text{emf}_{ref} = 5.278 \text{ mV}$$

and we have a single data point at T = 100.17°C where $(\text{emf}_{ref} - \text{emf}_{meas}) = -0.032$ mV

A second calibration point is known, since, at 0°C $\text{emf}_{meas} = 0$. Assuming linear behavior for the error between these two points yields the calibration curve shown above. The curve is used to correct a measured emf to an equivalent NIST standard thermocouple output.

b) (Note: part b of this problem requires using some judgement in setting uncertainty levels for various contributions to uncertainty)
One contribution to the uncertainty would result from the measured barometric pressure (at the design stage)

$$u_T = \left(\frac{\partial T}{\partial P}\right) u_P$$

Assume $u_P = \pm 0.05$ in. Hg and from

$$T_{st} = 212 + 50.422\left(\frac{P}{P_o} - 1\right) - 20.95\left(\frac{P}{P_o} - 1\right)^2$$

$$\frac{\partial T_{st}}{\partial P} = \frac{50.422}{P_o} - \frac{2(20.95)}{P_o}\left(\frac{P}{P_o} - 1\right)$$

at 30.1 in. Hg

$$\frac{\partial T_{st}}{\partial P} = 1.677 \text{ °F/in.Hg} = 0.93 \text{ °C/in. Hg}$$

The contribution to uncertainty would be

$$u_T = (0.93°C / \text{in. Hg})(0.05 \text{ in. Hg}) = \pm 0.047°C$$

Some estimate of the uncertainty due to the assumed behavior of $\text{emf}_{ref} - \text{emf}_{meas}$ must be made. A reasonable estimate may be to take 1/4 of the maximum deviation over the range at the midpoint such that in this case at 2.5 mV the uncertainty would be

$$\pm \frac{8 \text{ mV}}{1000}$$

yielding ±0.00044°C

COMMENT:

Without additional measured data points, a reasonable estimate of the deviation from the assumed linear behavior for $\text{emf}_{ref} - \text{emf}_{meas}$ yields an uncertainty estimate. Engineering judgement is required in applying this estimate for decisions in interpreting measured data or in measurement system design or selection.

PROBLEM 8.18

KNOWN: A J-type thermocouple is calibrated against an RTD, yielding calibration data over a range from 0°C to 100°C. The uncertainty in determining temperature using the RTD is ±0.01°C over the range 0 to 200°C

FIND:
   a) a polynomial to relate temperature and emf
   b) the uncertainty in a measured temperature using the system as calibrated
   c) the uncertainty in measured temperature using a specified indicator

ASSUMPTIONS: The calibration polynomial curve will be employed in data reduction

SOLUTION:
First, second, and third order polynomials for this data are
$$y = 0.34058 + 18.833x$$
$$y = 0.10989 + 19.157x - 0.06075x^2$$
$$y = -0.079403 + 19.926x - 0.45169x^2 + 0.048639x^3$$

Choice of an appropriate polynomial can be made for a particular application, depending upon the required uncertainty level. The standard error of the fit for the third order polynomial is 0.34, and for the fourth order is 0.46.

b) Error contributions are
$$\text{RTD - } \pm 0.01°C$$
$$\text{Poteniometer - } \sqrt{0.001^2 + 0.015^2} = \pm 0.015 \text{ mV}$$

which is equivalent to a temperature uncertainty of
$$0.015 \text{ mV} \left(\frac{1}{0.055 \text{ mV}/°C} \right) = \pm 0.27°C$$

and the value of s_e taken to be 0.34, yielding
$$u_T = \sqrt{0.01^2 + 0.27^2 + 0.34^2} = \pm 0.43°C$$

c) The readout uncertainty can be substituted for the potentiometer value and
$$u_T = \sqrt{0.01^2 + 0.32^2 + 0.34^2} = \pm 0.47°C$$

PROBLEM 8.19

KNOWN: A thermocouple is placed in a moving gas stream with

$U = 200$ ft/sec $c_p = 0.6$ Btu/lb$_m$ R

$h = 30$ Btu/hr-ft^2-R $T_s = 1200$ R

$T_p = 1400$ R $r = 0.22$

$F = 1$ $\varepsilon = 1$

FIND: a) T_∞

b) e_r

SOLUTION:

a) the static tempeature of the gas may be found from (8.36)

$$T_\infty = T_p - \frac{rU^2}{2g_c c_p}$$

which yields

$$T_\infty = 1400 - \frac{(0.22)(200)^2}{2(32.174)(0.6)(778)}$$

$$= 1400 \text{ R} - 0.293 \text{ R} = 1399.78 \text{ R}$$

b) The radiation error may be found from (8.30)

$$hA_s(T_\infty - T_p) = FA_s \varepsilon \sigma (T_p^4 - T_s^4)$$

$(30 \text{ Btu/hr-ft}^2\text{-R})(T_\infty - T_p) = 0.1714 \times 10^{-8} \frac{\text{Btu}}{\text{hr-ft}^2\text{-R}^4}(1400^4 - 1200^4)\text{R}^4$

$T_\infty = 1501.0$ R

$e_r = -101$ R

PROBLEM 8.20

KNOWN: The static temperature of air outside an aircraft is to be measured.

U = 300 mph = 438.3 ft/sec

Altitude = 20,000 ft

r = 0.75

T_∞ = 413 R

c_p = 0.24 Btu/lb_m-R

ρ_{air} = 0.0442 lb_m/ft^3

FIND: T_p

SOLUTION:

From (8.36)

$$T_p = T_\infty + \frac{rU^2}{2g_c c_p}$$

$$= 413 + \frac{0.75(438.3^2) \, ft^2/sec^2}{2\left(32.174 \frac{ft \cdot lb_m}{lb \cdot sec^2}\right)\left(0.24 \frac{Btu}{lb_m R}\right)\left(778 \frac{ft \cdot lb}{Btu}\right)}$$

which yields

T_p = 425 R

PROBLEM 8.21

KNOWN: A sheathed thermocouple, as shown in Figure 8.38.

FIND: An estimate of the upper limit for conduction error for such a probe.

SOLUTION: From (8.27)

$$e_c = \frac{T_w - T_\infty}{\cosh mL}$$

where

$$mL = \sqrt{hP/kA}\, L$$

For immersion depth as a parameter, an estimate of the conduction error requires a model of the effective thermal conductivity of the thermocouple probe. A conservative estimate for many constructions could be an average of the thermocouple, sheath and insulating materials. Consider the following values:

$k_{constantan}$ = 23 W/m-K
$k_{stainless}$ = 15 W/m-K
k_{insul} = 0.05 W/m-K

Averaging these values yields 12.7 W/m-K. Considering the thermocouple probe to be cylindrical in shape,

$$mL = \sqrt{\frac{4h}{k_{eff} D}}\, L$$

For a probe having a diameter of 0.25 cm and an immersion length of 5 cm, the conduction error is plotted as a function of h in the figure below.

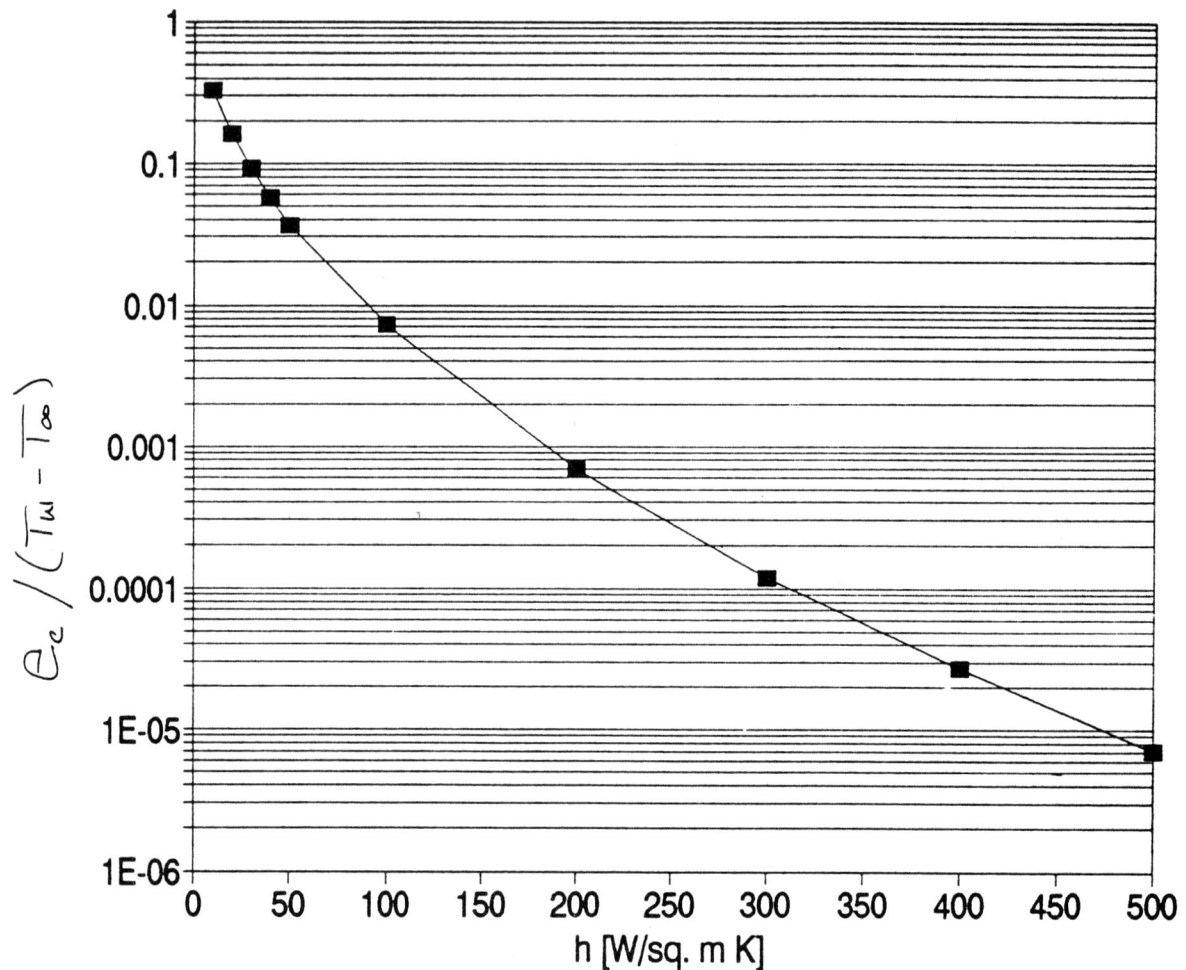

PROBLEM 8.22

KNOWN: An iron-constantan thermocouple is placed in a moving gas stream (as shown in Figure 8.39)

$T_{ref} = 100°C$ $T_w = 260°C$

$h = 70$ Btu/hr-ft^2-°F $V = 200$ ft/sec

emf $= 14.143$ mV $r = 0.7$

$c_p = 0.24$ Btu/lb$_m$°F $\varepsilon = 0.25$

FIND:
a) T_p

b) e_r and e_u

ASSUMPTIONS: Radiation and velocity errors are additive

SOLUTION:

a) From Table 8.6, $T_p = 355.8°C$

b) The velocity error is given by

$$e_U = T_p - T_\infty = \frac{rU^2}{2g_c c_p} = \frac{0.7(200)^2}{2(32.174)(0.24)(778)} = 2.33 R$$

and the radiation error by

$$e_r = T_p - T_\infty = \frac{\varepsilon\sigma}{h}\left(T_w^4 - T_p^4\right) = \frac{0.25(0.1714 \times 10^{-8})}{70}\left(960^4 - 1132.4^4\right) = -4.87 R$$

The total error is then estimated as

$$e = e_U + e_r = 2.33 + (-4.87) = -2.54 R = -1.4°C$$

PROBLEM 8.23

KNOWN: $E_i = 1.564$ V At 125°C, from Example 8.5, $B_{RT} = 247\Omega$

FIND: Show that $B_{RT} = 247\Omega$, and determine the values of B_{RT} at 150°C and 100°C

SOLUTION:

Since

$$B_{R_T} = \sqrt{\left[\frac{\partial R_T}{\partial R_1}(B)_{R_1}\right]^2 + \left[\frac{\partial R_T}{\partial E_i}(B)_{E_i}\right]^2 + \left[\frac{\partial R_T}{\partial E_1}(B)_{E_1}\right]^2}$$

and

$$(B)_{R_1} = \pm 1.96 \text{ k}\Omega \quad (B)_{Ei} = \pm 1.96 \text{ k}\Omega$$

The sensitivity indices are functions of temperature with

$$R_T = R_1\left(\frac{E_i}{E_1} - 1\right)$$

and

Sensitivity Indices	100°C	125°C	150°C
$\frac{\partial R_T}{\partial R_1} = \left(\frac{E_i}{E_1} - 1\right)$	0.042	0.022	0.012
$\frac{\partial R_T}{\partial E_i} = \frac{R_1}{E_1}$	86942	85238	84466
$\frac{\partial R_T}{\partial E_1} = \frac{-R_1 E_i}{E_1^2}$	90591	87076	85505

This yields values of B_{RT} of

T (°C)	B_{RT} (Ω)
100	264
125	247
150	241

PROBLEM 8.24

KNOWN: A thermocouple circuit emf is measured by a potentiometer having limits of error as

 0.05% of reading + 15 μV at 25°C
 and a resolution of 5 μV.

The connecting block temperature is 21.5 ± 0.2°C
and the potentiometer junctions are 25 ± 0.2°C.

FIND: T_1

SOLUTION:
The error sources for the potentiometer may be combined,

$$u_{emf} = \sqrt{\left(\frac{0.05}{100} 9000 + 15\right)^2 + 2.5^2}$$

$$u_{emf} = \pm 19.66 \text{ μV} = \pm 0.020 \text{ mV}$$

Then since

$$u_T = \left(\frac{\partial T}{\partial emf}\right) u_{emf}$$

and the sensitivity of the thermocouple is 0.055 mV/°C (from Table 8.6)

$$u_T = \left(\frac{1°C}{0.055 \text{ mV}}\right)(0.020 \text{ mV}) = \pm 0.366°C$$

The contribution from the uncertainty in the reference junction at the potentiometer is ±0.2°C, and the limits of error on the thermocouple are ±2.2°C. Thus the total uncertainty in temperature is

$$u_T = \sqrt{(0.366)^2 + (0.2)^2 + (2.2)^2} = \pm 2.24°C$$

The emf referenced to 0°C would be

 $emf_{0-T} = 9 \text{ mV} + 1.096 \text{ mV} = 10.096 \text{ mV}$

yielding

 $T = 187.7 \pm 2.24°C$.

PROBLEM 8.25

KNOWN: A concentration of salt of 600 ppm in tap water will cause a 0.1°F change in the freezing point.

FIND: Error in ice bath temperature having 1500 ppm of salt.

SOLUTION:
Consider two error sources for this ice bath,
1. salt $\pm 0.25°F$
2. local temperature variations $\pm 0.1°F$

The resulting design stage uncertainty may be found as

$$u_T = \sqrt{(0.25)^2 + (0.1)^2} = \pm 0.27°F$$
$$T = 32.0 \pm 0.27°F$$

PROBLEM 8.26

KNOWN: An RTD is to be calibrated; the RTD forms one leg of a Wheatstone bridge, and has

$$\alpha = 0.00392°C^{-1} \pm 1 \times 10^{-5} \ (95\%)$$

$$u_R = \pm 0.001 \ \Omega \ (95\%).$$

At balanced conditions with T = 0°C, R_c = 100.000 Ω and at 100°C, R_c = 139.200 Ω.

FIND: R_{RTD} at 0°C and 100°C, and the uncertainty at the design stage at these temperatures.

SOLUTION:

At balanced conditions

$$\frac{R_{RTD}}{R_a} = \frac{R_c}{R_b}$$

Thus

$$R_{RTD} = R_c$$

at 0°C R_{RTD} = 100.000 Ω

at 100°C R_{RTD} = 139.200 Ω

Expressing the relationship between temperature and resistance as

$$T = \frac{1}{\alpha}\left(\frac{R_{RTD}}{R_o} - 1\right) + T_o$$

the uncertainty at the design stage may be expressed, with $\gamma = \dfrac{R_{RTD}}{R_o}$

$$u_T = \left[\left(\frac{\partial T}{\partial \alpha} u_\alpha\right)^2 + \left(\frac{\partial T}{\partial \gamma} u_\gamma\right)^2\right]^{1/2}$$

The sensitivities are found as

$$\frac{\partial T}{\partial \alpha} = \frac{1-\gamma}{\alpha^2} = \left(\frac{1-3}{0.00392^2}\right) = -130154°C^2$$

$$\frac{\partial T}{\partial \gamma} = \frac{1}{\alpha} = 255°C$$

at R_c = 300 Ω
$R_{RTD} = R_c$ which implies R_{RTD} = 300 Ω, R_o = 100 Ω γ = 3

We must estimate the uncertainty in γ. Since

$$R_{RTD} = \frac{R_a}{R_b} R_c$$

with $u_R = \pm 0.001\ \Omega$

$$u_{R_{RTD}} = (3 \times 0.001) = \pm 0.003\ \Omega$$

and

$$u_\gamma = \left[\left(\frac{1}{R_o} u_{R_{RTD}} \right)^2 + \left(\frac{-R_{RTD}}{R_o^2} u_{R_o} \right)^2 \right]^{1/2}$$

$$= \left[\left(\frac{1}{100} 0.003 \right)^2 + \left(\frac{-300}{100^2} 0.001 \right)^2 \right]^{1/2} = \pm 0.0055$$

yielding

$$u_T = \left[(-130154 \times 10^{-5})^2 + (255 \times 0.0055)^2 \right]^{1/2} = \pm 1.91°C$$

PROBLEM 8.27

KNOWN: A T-type thermopile is used to measure the temperature difference to establish heat flux across an insulation. The pertinent variables and their values are:

$A_c = 15 \text{ m}^2$ $\quad k = 0.4 \text{ W/m-K}$
$L = 0.25 \text{ m}$ $\quad \Delta T = 5°C$

The uncertainty in the measured emf is ± 0.04 mV.

FIND: The number of junctions in the thermopile to yield an uncertainty level of 5% in the heat flux across the insulation.

ASSUMPTIONS: NIST standard emf versus temperature relationship, and an average temperature in the insulation of 40°C.

SOLUTION:
The heat flux is expressed as

$$Q = kA_c \frac{\Delta T}{L}$$

For the purposes of the present analysis, express the uncertainty in Q as a percentage, yielding

$$\frac{u_Q}{Q} = \frac{u_{\Delta T}}{\Delta T}$$

To determine the uncertainty in ΔT, the sensitivity to the uncertainty in emf must be detemined. From the equation in Table 8.7, the value can be determined using

$$\frac{dE}{dT} = c_1 + c_2 T + c_3 T^2 + c_4 T^3 + c_5 T^4 + c_6 T^5 + c_7 T^6 + c_8 T^7$$

This expression yields a value of 0.042 mV/°C. Thus

$$u_{\Delta T} = \frac{1}{N(0.042)} u_{emf}$$

where

$u_{emf} = \pm 0.04$ mV

Then with

$$\frac{u_{\Delta T}}{\Delta T} = 5\% = \frac{1}{5N}$$

which yields N = 4

PROBLEM 8.28

KNOWN: A T-type thermocouple referenced to 0°C is used to measure 100°C

FIND: The output emf.

ASSUMPTIONS: NIST standard emf versus temperature relationship.

SOLUTION:

From Table 8.7, the polynomial expression for emf as a function of temperature yields an emf of 4.2785 mV at 100°C.

PROBLEM 8.29

KNOWN: A T-type thermocouple referenced to 0°C has an output of 1.2 mV.

FIND: The temperature of the measuring junction.

ASSUMPTIONS: NIST standard emf versus temperature relationship.

SOLUTION:

From Table 8.7, the polynomial expression for emf as a function of temperature yields an temperature of 30.086°C for an emf of 1200 μV.

PROBLEM 8.30

KNOWN: A T-type thermocouple and voltmeter form a temperature measuring system. The temperature at the voltmeter is 25°C, and the output emf is 10 mV.

FIND: The measuring junction temperature.

SOLUTION:

The law of intermediate temperatures allows the following superposition to be used to establish an equivalent emf referenced to 0°C.

$$emf_{25-T} + emf_{0-25} = emf_{0-T}$$

From Table 8.7 the polynomial equation for E = f(T) yields

$$emf_{0-25} = 992 \ \mu V$$

and
$$emf_{0-T} = 992 + 10,000 = 10,992 \ \mu V$$

which yields for T, from Table 8.7, a value of 231.542°C.

COMMENT: A calculator or mathematical analysis software is almost essential to solve for a temperature from the polynomial expression in Table 8.7. NIST publications are also available which contain tables of emf as a function of temperature for a variety of thermocouple types.

PROBLEM 9.1

FIND: Convert between units of pressure

SOLUTION

Absolute pressure reference scale:

$$1 \text{ atm abs} = 14.69 \text{ psia} = 101.325 \text{ kPa abs} = 101{,}325 \text{ N/m}^2 \text{ abs}$$
$$1 \text{ atm abs} = 760 \text{ mm Hg abs} = 406 \text{ in H}_2\text{O abs}$$

Conversion factors:

$$14.69 \text{ psi} = 101.325 \text{ kPa} = 101{,}325 \text{ N/m}^2$$
$$14.69 \text{ psi} = 1 \text{ atm} = 29.92 \text{ in Hg} = 406 \text{ in H}_2\text{O} = 760 \text{ mm Hg}$$

(a) 10.8 psia x 101325 N/m^2/14.69 psi = 74 442 N/m^2 abs = 74.442 kPa abs
 10.8 psia x 1 atm/14.69 psi = 0.73 atm abs
 10.8 psia x 760 mm Hg/14.69 psi = 559 mm Hg abs
 10.8 psia x 406 in H$_2$O/14.69 psi = 298 in H$_2$O abs

(b) 1.75 atm abs x 101325 N/m^2/1 atm = 177,319 N/m^2 = 177.319 kPa abs
 1.75 atm abs x 14.69 psi/1 atm = 25.71 psia
 1.75 atm abs x 760 mm Hg/1 atm = 1329.94 mm Hg abs
 1.75 atm abs x 406 in H$_2$O/1 atm = 710.5 in H$_2$O abs

(c) 30.36 in H$_2$O abs x 101,325 N/m^2/406 in H$_2$O = 7,577 N/m^2 abs
 = 7.58 kPa abs
 30.36 in H$_2$O abs x 14.69 psi/406 in H$_2$O = 1.098 psia
 30.36 in H$_2$O abs x 760 mm Hg/406 in H$_2$O = 56.83 mm Hg abs
 30.36 in H$_2$O abs x 1 atm/406 in H$_2$O = 0.075 atm abs

(d) 791 mm Hg abs x 406 in H$_2$O/760 mm Hg = 422 in H$_2$O abs
 791 mm Hg abs x 1 atm/760 mm Hg = 1.04 atm abs
 791 mm Hg abs x 101,325 N/m^2/760mm Hg = 105,458 N/m^2 abs
 = 105.458 kPa abs
 791 mm Hg abs x 14.69 psi/760 mm Hg = 15.30 psia

COMMENT

Because $p_{gauge} = p_{abs} - p_{ref}$ all of the above pressures are easily converted to a gauge pressure, referenced to standard atmospheric pressure, by subtracting the equivalent of 1 atm abs from each absolute pressure. For example, 10.80 psia is converted to a gauge pressure referenced to standard atmospheric pressure by subtracting 14.69 psia: 10.8 psia - 14.69 psia = - 3.89 psi. This gauge pressure can then be easily converted to other units of gauge pressure by using the conversion factors.

PROBLEM 9.2

FIND: Convert into absolute pressure

SOLUTION

$$P_{abs} = P_{gauge} + P_{ref} \quad \text{where here: } P_{ref} = 1 \text{ atm abs.}$$

Absolute pressure reference scale:

1 atm abs = 14.69 psia = 101.325 kPa abs = 101,325 N/m² abs
1 atm abs = 760 mm Hg abs = 406 in H₂O abs

Conversion factors:

14.69 psi = 101.325 kPa = 101,325 N/m²
14.69 psi = 1 atm = 29.92 in Hg = 406 in H₂O = 760 mm Hg

(a) -0.55 psi + 14.69 psia = 14.14 psia = 0.963 atm abs

(b) 100 μm Hg + 760 mm Hg abs (1 mm/1000 μm) = 760.101 mm Hg abs
  = 1.0001 atm abs

(Note: units of μm Hg are generally used only in absolute terms so as to describe very low pressures. The value of 100 μm Hg abs would be a very low pressure (equivalent to 1 x 10⁻⁴ atm abs) such as might be found in an nearly evacuated vessel. But 100 μm Hg represents a value nearly at atmospheric pressure.)

(c) 98.6 kPa + 101.325 kPa abs = 199.925 kPa abs = 1.97 atm abs

(d) 3 in H₂O + 406 in H₂O abs = 409 in H₂O abs = 1.007 atm abs

COMMENT

Inspection of this problem should reveal that gauge pressures may be positive or negative in value depending on whether they represent above atmospheric pressure or below atmospheric pressure values. Absolute pressures are positive numbers because they are referenced to absolute zero pressure.

PROBLEM 9.3

KNOWN: $H = 20$ in H_2O
$P_{atm} = 14.59$ psia
$\gamma_{H2O} = 62.4$ lb/ft^3

FIND: The tank pressure p_1.

PROPERTIES: $\gamma_{H2O} = 62.4$ lb/ft^3

SOLUTION

Referring to the sketch below,

$p_1 = P_{atm} + \gamma_{H2O} H$

$= 14.59$ psia $+ (62.4$ lb/ft$^3)(20/12$ ft$)(1$ ft$^2/144$ in$^2) = 15.31$ psia

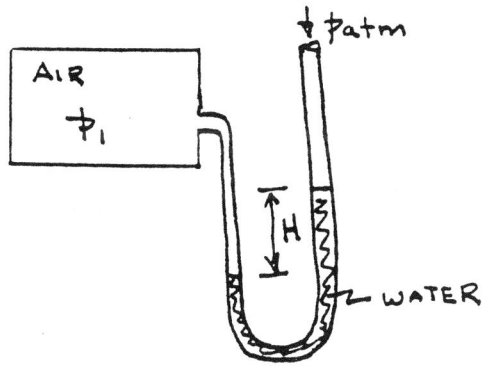

PROBLEM 9.4

KNOWN: Deadweight tester provides a calibration pressure, p_c
Tester:
   $W = 11.5$ lb
   $A_p = 0.785$ in^2
   $W_p = 2.43$ lb
   $p_{amb} = 30.15$ in Hg abs $= 14.80$ psia $= 102.104$ kPa abs
   $z = 59$ feet (sea level datum)
   $\phi = 42°$

FIND: p_c

PROPERTIES: $\gamma_{air} = 0.075$ lb/ft^3
   $\gamma_{ss} = \gamma_{mass} = 78.4$ kN/m$^3 = 488$ lb/ft^3

SOLUTION

Referring to the free-body diagram,

$$\Sigma F_y = 0 = p_i A_p + F_B - 11.5 \text{ lb} - 2.43 \text{ lb} - p_{amb} A_p$$

where F_B is the buoyancy force. Neglecting F_B, the indicated pressure is

$$p_i = 17.745 \text{ psi}$$

Now, the actual pressure provided by the deadweight tester is estimated by

$$p_c = p_i(1 + e_1 + e_2)$$

where e_1 provides a correction for altitude effects and e_2 provides the correction for the neglected buoyancy effects. From (9.6a),

$$e_1 = -0.0003$$

and from (9.9),

$$e_2 = -\gamma_{air}/\gamma_{masses} = -(0.075 \text{ lb/ft}^3/488 \text{ lb/ft}^3) = -0.0002$$

Then, the actual deadweight pressure is estimated by

$$p_c = 17.745 \text{ psi } (1 - 0.0003 - 0.0002) = 17.727 \text{ psi}$$
$$= 32.530 \text{ psia} = 122.267 \text{ kPa abs}$$

COMMENT

The inclusion of the correction factors e_1 and e_2 is an example of correcting a measurement for known bias error. The values obtained for e_1 and e_2 are only estimates of the actual bias error introduced by elevation and buoyancy effects. Not only are the correction equations approximate, the values used in them, for example the specific weight terms, can only be estimated. The correction reduces the bias error in p_c brought about by these two effects, but it does not eliminate it.

PROBLEM 9.5

KNOWN: Inclined tube manometer
$\Delta L = 5.6$ cm H_2O
$\theta = 30^0$

FIND: ΔH

SOLUTION

Referring to Figure 9.7, for an inclined manometer,

$$\Delta H = \Delta L \sin \theta$$

Further, this deflection away from a null balance condition is referenced to the pressure on the open end of the tube. So the device measures a gauge (referenced to local atmosphere pressure) or differential pressure (referenced to some other pressure).

$$\Delta H = 5.6 \text{ cm } H_2O \times \sin 30^0$$

$$= 2.8 \text{ cm } H_2O$$

COMMENT

Again referring to Figure 9.7, if the manometer fluid deflects towards the tank, the pressure applied to the tank is below the reference pressure ($p_2 < p_1$). If the deflection is away from the tank, the pressure applied to the tank is above the reference pressure ($p_2 > p_1$).

PROBLEM 9.6

FIND: Compare K for inclined and U-tube manometers.

SOLUTION

For a U-tube manometer, H is the measured output,

$$\Delta p = (\gamma_m - \gamma)H \text{ so}$$
$$K = dH/d(\Delta p) = 1/(\gamma_m - \gamma)$$

For an inclined manometer, $H = L\sin\theta$ where L is the measured output.

$$\Delta p = (\gamma_m - \gamma)L\sin\theta \text{ so}$$
$$K = dL/d(\Delta p) = 1/[(\gamma_m - \gamma)\sin\theta]$$

PROBLEM 9.7

KNOWN: Inclined manometer measures air using mercury as its fluid.
$\theta = 30°$

FIND: K

SOLUTION

For an inclined manometer, where L is the measured output

$\Delta p = (\gamma_m - \gamma)H = (\gamma_m - \gamma)L\sin\theta$ so
$K = dL/d(\Delta p) = 1/[(\gamma_m - \gamma)\sin\theta]$

$= 1/\{[(13.6)(9800 \text{ N/m}^3) - 11 \text{ N/m}^3)]\sin 30°\} = 0.015 \text{ mm/N/m}^2$

PROBLEM 9.8

KNOWN: Conditions of Example 9.2. $\theta \rightarrow 90^\circ$

FIND: $(u_d)_p$

SOLUTION

From Example 9.2:

$$(u_d)_p = \pm[(u_{\gamma m}L\sin\theta)^2 + (\{\gamma_m - \gamma\}u_L\sin\theta)^2 + (L\{\gamma_m - \gamma\}u_\theta\cos\theta)^2]^{.5}$$

For a U-tube manometer at the stated conditions, $L \approx 10.25$ mm and $\theta = 90^\circ$. Then with $\gamma_m = 9770$ N/m^3; $\gamma = 11.5$ N/m^3; $u_{\gamma m} = 49$ N/m^3; $u_L = 0.0007$ m

$$(u_d)_p = \pm[.5^2 + 6.8^2 + 0]^{.5} = \pm 6.82 \text{ N/m}^2$$

COMMENT

The uncertainty in measured pressure increases nearly 50% by going from an inclined manometer with $\theta = 30^\circ$ to a U-tube manometer when operating at these pressures.

PROBLEM 9.9

KNOWN: Steel diaphragm
$t = 0.1$ in
$E_m = 30 \times 10^6$ psi
$d = 2r = 0.75$ in
$\rho = 0.28$ lb$_m$/in^3
$\nu_p = 0.32$

FIND: $y_{max}, \omega_n, \Delta p_{max}$

SOLUTION

The maximum elastic deflection of a metallic diaphragm is about one third of the diaphragm thickness,

$$y_{max} \approx t/3 = 0.033 \text{ in} = 0.85 \text{ mm}$$

The natural frequency can be computed directly from (9.10),

$$\omega_n = 64.15 \left(\frac{(30 \times 10^6 \text{ psi})(0.1 \text{ in})^2 (386 \text{ lb}_m\text{-in}/\text{lb-s}^2)}{12 (1 - .32^2)(0.375 \text{ in})^4 (0.28 \text{ lb}_m/\text{in}^4)} \right)^{1/2}$$

$$= 2.8 \times 10^6 \text{ rad/s} \text{ or } 450 \text{ kHz}.$$

The maximum differential pressure which can be applied across a diaphragm is limited in part by y_{max}. Rearranging (8.11) with $y_{max} = t/3$ gives

$$\Delta p_{max} = \frac{(16)(30 \times 10^6 \text{ lb/in}^2)(0.1 \text{ in})^4}{9 (1 - 0.32^2)(0.375 \text{ in})^4} =$$

$$= 300{,}460 \text{ psi} = 2.07 \text{ GPa}$$

COMMENT

These relatively high numbers are due to the relatively thick and small diameter steel diaphragm used. The numbers are not unusual for high pressure diaphragm tranducers. Some transducers on the market permit the user to interchange diaphragms of different thicknesses to change the maximum pressure differential range allowed. This is a cost saving feature for the user.

PROBLEM 9.10

KNOWN: Strain gauge, diaphragm tranducer
Δp = 1, 10, 100 psi
Transducer:
  Accuracy: within 0.1% of reading (i.e. $u_{\Delta p}/\Delta p = 0.001$)
Voltmeter:
  Resolution: $10\mu V$
  Accuracy: within 0.1% reading

FIND: u_d in indicated pressure

ASSUMPTIONS: Transducer: $K_t = 1$ V/psi
Voltmeter: $K_E = 1$ V/V

SOLUTION

Transducer:

$$u_{\Delta p} = 0.001\, \Delta p$$

Voltmeter: (Note the inclusion of K_E for scaling and units)

$$u_E = [u_o^2 + u_c^2]^{0.5} = [(5 \times 10^{-6})^2 + (0.001 K_E E)^2]^{0.5}$$

System: (Note the inclusion of K_t for scaling and units)

$$u_d = \pm[u_{\Delta p}^2 + (u_E/K_t)^2]^{0.5}$$

$$= \pm[(0.001\, \Delta p)^2 + (\{(5 \times 10^{-6})^2 + (0.001 K_E E)^2\}^{0.5}/K_t)^2]^{0.5}$$

With $K_t = 1$ V/psi and $K_E = 1$ V/V,

Δp [psi]	E_o [V]	u_d [psi]
1	1	0.0014
10	10	0.0141
100	100	0.1410

COMMENT

In all cases above, the transducer and output instrument errors dominate over the small resolution error.

By altering the sensitivities one can alter the magnitude of the resulting uncertainties.

PROBLEM 9.11

KNOWN: U-tube manometer is used to measure gas pressure.
  $P_{gas} \leq 10$ psi
  Several manometric fluids available: oil, water, mercury

FIND: Choose an appropriate manometric fluid.

PROPERTIES: water: $S = 1$; $\gamma_{h2o} = 62.4$ lb/ft^3 (given)
  mercury: $S = 13.57$ (given); $\gamma = S\gamma_{h2o}$
  oil: $S = 0.82$ (given); $\gamma = S\gamma_{h2o}$

ASSUMPTIONS: The specific weight of a gas is negligible relative to that of the manometric fluids.

SOLUTION

To select an appropriate fluid, we must consider at least two things. (1) The manometric fluid should not be soluble with the working fluid, and (2) the manometer deflection should be of a reasonable magnitude.

Referring to the sketch, the relation between pressure and manometric fluid deflection is given by,

$$p - p_{atm} = \gamma H + \gamma_{air} L_1 - \gamma_{gas}(H + L_2)$$

Neglecting the effects of the gases,

$$p - p_{atm} = \Delta p = \gamma H$$

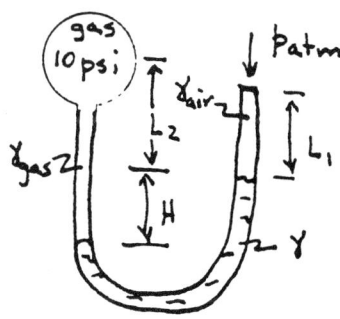

Water:

$$H = \Delta p/\gamma = (10 \text{ psi})(144 \text{ in}^2/\text{ft}^2)/62.4 \text{ lb/ft}^3 = 23 \text{ feet}$$

Oil:

$$H = \Delta p/\gamma = (10 \text{ psi})(144 \text{ in}^2/\text{ft}^2)/(0.82)(62.4 \text{ lb/ft}^3) = 28.1 \text{ feet}$$

Mercury:

$$H = \Delta p/\gamma = (10 \text{ psi})(144 \text{ in}^2/\text{ft}^2)/(13.57)(62.4 \text{ lb/ft}^3) = 1.7 \text{ feet}$$

While sensitivity will always favor the lighter fluid, the logistics of this application clearly suggest that mercury will be the appropriate choice.

PROBLEM 9.12

KNOWN: Air pressure to be measured using either a mercury filled U-tube manometer or inclined tube manometer.
$200 \leq p \leq 400$ N/m^2
T = 20°C
U-tube Manometer:
    Resolution: 1 mm; Zero error: 0.5 mm
Inclined-tube manometer
    Resolution: 1 mm; Zero error: 0.5 mm;
    Inclination angle: 30° ± 0.5°

FIND: u_d in equivalent head pressure measured by either manometer

PROPERTIES: mercury: S = 13.57; water: γ = 9780 N/m^3

SOLUTION

U-tube manometer:

Because the measured deflection is the equivalent head pressure, the uncertainty in equivalent head pressure at the design-stage will be due only to the ability to measure the deflection at a given pressure.

$$u_d = \pm(u_0^2 + u_c^2)^{0.5}$$

If we assume that u_c is based only on the zero error given,

$$u_0 = 0.5 \text{ mm} \qquad u_c = 0.5 \text{ mm}$$

$$u_d = \pm(0.5^2 + 0.5^2)^{0.5} = \pm 0.71 \text{ mm} \quad (95\%)$$

This result is independent of pressure.

Inclined-tube manometer

The equivalent head pressure is related to the manometer deflection by,

$$H = L \sin \theta$$

where L is the measured deflection. We can estimate L:

$$L = H/\sin\theta = \Delta p/\gamma\sin\theta$$

At 200 N/m², L = 41 mm. At 400 N/m², L = 82 mm.

For the inclined manometer, the uncertainty in the manometer deflection, L, at the design stage is

$$(u_d)_L = \pm(u_0^2 + u_c^2)^{0.5} = \pm(0.5^2 + 0.5^2)^{0.5} = \pm 0.71 \text{ mm}$$

where u_0 = 0.5 mm and u_c = 0.5 mm as before. However, the uncertainty in equivalent head, H, depends on the uncertainty in two variables,

$$H = f(L, \theta)$$

$$(u_d)_H = \pm\left[\left(\frac{\partial H}{\partial L} u_L\right)^2 + \left(\frac{\partial H}{\partial \theta} u_\theta\right)^2\right]^{1/2}$$

$$= \pm\left[(\sin\theta \, u_L)^2 + (L\cos\theta \, u_\theta)^2\right]^{1/2}$$

We set $u_\theta = 0.5° = 0.0087$ rad from the problem statement.

At 200 N/m²:

$$(u_d)_H = \pm[\{(0.5)(0.71)\}^2 + \{(41)(0.87)(0.0087)\}^2]^{0.5} = 0.47 \text{ mm} \quad (95\%)$$

At 400 N/m²:

$$(u_d)_H = \pm[\{(0.5)(0.71)\}^2 + \{(82)(0.87)(0.0087)\}^2]^{0.5} = 0.72 \text{ mm} \quad (95\%)$$

COMMENT

At the lower pressure, the uncertainty is reduced by a factor of about sin θ by using the inclined manometer. But at higher pressures, the uncertainty in θ becomes increasingly important. In this application, the uncertainty in θ cancels out the benefits of the better sensitivity of the inclined instrument.

PROBLEM 9.13

KNOWN: A water filled inclined-tube manometer.
Inclination angle is variable.
$\Delta p \approx 10{,}000 \text{ N/m}^2$
$T = 20°C$
Manometer:
Resolution: 1 mm; Zero error: 0.5 mm; Inclination error: $\pm 1°$

FIND: u_d in pressure as a function of θ

PROPERTIES: water: $\gamma_m = 9770 \text{ N/m}^3$

ASSUMPTIONS: $(u_{\gamma m}/\gamma_m) = 0.5\%$. Neglect effects of the ambient air.

SOLUTION

$$\Delta p = L(\gamma_m - \gamma)\sin\theta \approx L\gamma_m \sin\theta$$

Then, $\Delta p = f(L, \gamma_m, \theta)$,

$$(u_d)_{\Delta p} = \pm\left[\left(\frac{\partial \Delta p}{\partial L} u_L\right)^2 + \left(\frac{\partial \Delta p}{\partial \gamma_m} u_{\gamma m}\right)^2 + \left(\frac{\partial \Delta p}{\partial \theta} u_\theta\right)^2\right]^{1/2}$$

$$= \pm\left[(\gamma_m \sin\theta \, u_L)^2 + (L\sin\theta \, u_{\gamma m})^2 + (L\gamma_m \cos\theta \, u_\theta)^2\right]^{1/2}$$

For the inclined manometer, the uncertainty in the manometer deflection, L, at the design stage is

$$(u_d)_L = \pm(u_0^2 + u_c^2)^{0.5} = \pm(0.5^2 + 0.5^2)^{0.5} = \pm 0.71 \text{ mm}$$

where $u_0 = 0.5$ mm and $u_c = 0.5$ mm.

$$(u_d)_\theta = 1° = 0.0175 \text{ rad}$$

$$(u_d)_{\gamma m} = (9770 \text{ N/m}^3)(0.005) = 49 \text{ N/m}^3$$

Then,

Inclination	L	$(u_d)_{\Delta p}$
	[m]	[N/m²]

Inclination	L [m]	$(u_d)_{\Delta p}$ [N/m²]	
10°	5.894	994	u_θ dominates
30°	2.047	307	u_θ dominates
60°	1.182	113	u_θ dominates
80°	1.039	59	u_γ, u_θ large
90°	1.024	50	u_γ dominates

COMMENT

At large pressures, u_θ becomes very important (compare with Problem 9.12). In practice, the inclination angle must be carefully set. But even so, the inclined manometer is selected for deflections up to only about H = 0.25 m.

PROBLEM 9.14

KNOWN: Capacitance pressure transducer of Figure 9.14.
$C_1 = 0.01 \pm 0.005 \, \mu F$
$E_i = 5 \pm 1\% \, V$
$A = 8 \pm 0.01 \, mm^2$
$t = 1.5 \pm 0.1 \, mm$
$\Delta t = 0.2 \, mm$

FIND: C and E_o

SOLUTION

From (9.12)

$$C = c\epsilon A/t$$

and from (9.13),

$$E_o = (C_1/C)E_i$$

At $t_o = 1.5$ mm

$$C = (0.0885)(1)(8 \, mm^2)/(1.5 \, mm)(10 \, mm/cm) = 0.0472 \, \mu F$$

and

$$E_o = (0.01/.0472)(5V) = 1.0593 \, V$$

At $t_o + \Delta t = 1.7$ mm

$$C = 0.0416 \, \mu F$$

$$E_o = (0.01/.0416)(5V) = 1.2006 \, V$$

PROBLEM 9.15

KNOWN: Diaphragm pressure transducer:
$$u_c = \pm 0.5 \text{ psi}$$
Voltmeter
$$u_c = \pm 10 \, \mu V$$
$$u_o = 1 \, \mu V$$
System (0 to 100 psi)
$$p = 0.564 + 24E \pm 1 \text{ psi} \quad (95\%) \quad \text{with } N = 5$$
Installation errors: $B = \pm 0.5$ psi

FIND: u_p

SOLUTION

The system sensitivity is: $K = dp/dE = 24$ psi/V. The system uncertainty includes,

Instrument errors, u_c:
  transducer uncertainty $u_t = \pm 0.5$ psi
  voltmeter uncertainty $u_E = (1 \times 10^{-6} \text{ V})(24 \text{ psi/V}) = \pm 2.4 \times 10^{-4}$ psi
Data reduction errors, u_1:
  curve fit uncertainty $u_{cf} = \pm 1$ psi
Installation effects, u_2:
  installation errors $u_{in} = \pm 0.5$ psi

So that: $u_p = \pm [.5^2 + (2.4 \times 10^{-4})^2 + 1^2 + .5^2]^{.5} = \pm 1.22$ psi (95%)

Alternatively:
$u_t = B_t$; $u_E = B_E$; $u_{in} = B_{in}$; $u_{cf} = P_{cf} = S_{yx}$ with $\nu = 4$ then,
$u_p = \pm[B^2 + (tP)^2]^{.5} = \pm 1.22$ psi (95%)

COMMENT

There is no magic in uncertainty analysis. Different approaches will yield similar (but not necessarily identical) results provided that the same errors are included. The important thing is to do an analysis!

PROBLEM 9.16

KNOWN: pressure transducer: $t_{90} = 10$ ms; $\omega_d = 200$ Hz; $\zeta = 0.8$

FIND: Test plan to verify given specifications. Estimate frequency response.

SOLUTION

This problem is open-ended an could form the basis of a lab exercise. Below some valid points appropriate to a test plan are made: (i) A step test should be developed to test for the rise time. An appropriate magnitude for the pressure rise is from atmospheric pressure to the expected pressure at top dead center for this engine (note: the compression ratio for an IC engine is about 8:1 or 9:1). Transducer output can be measured on a storage oscilloscope or suitable data acquisition system.

For a damping ratio of 0.8, the transducer will still demonstrate a modest ringing during a step test. But it will require excellent resolution in the measuring device to observe an accurately measure. With adequate resolution, the maximum amplitudes in the oscillation can be plotted versus time to estimate $\omega_n\zeta$ (note: this will be the slope of the line if plotted on semi-log axes). The period of oscillation will be related to $\omega_d = \omega_n(1-\zeta^2)^{.5}$. The damping ratio and natural frequency are found by solving these two pieces of information simultaneously.

(ii) Car: 4-cylinder at 5000 RPM = 83 rps. This suggests pressure changes of about 83/4 ≈ 21 Hz. For the transducer,
$$f_n = f_d/(1-\zeta^2)^{.5} = 200 \text{ Hz}/(1-.8^2)^{.5} = 333 \text{ Hz Hence,}$$
$f/f_n \approx 0.063$ and $M(21 \text{ Hz}) \approx 1$. So yes.

PROBLEM 9.17

KNOWN: Steel diaphragm transducer: $t = 0.001$ m; $r = 0.003$ m

FIND: ω_n, p_{max}

SOLUTION

The natural frequency is given by (9.10):

$$\omega_n = (64.15)([200 \times 10^9 \text{N/m}^2][.001\text{m}]^2/[12][1-0.35^2][.003\text{m}]^4[7832 \text{kg/m}^3])^{.5}$$

$$= 173{,}000 \text{ r/s} \quad \text{or} \quad f_n = 127 \text{ kHz}$$

The maximum elastic displacement of the diaphragm is limited to about $t/3$. Rearranging (9.11) gives:

$$\Delta p = 16 E_m t^3 (t/3) / 3(1-\nu_p^2) r^4$$

For steel with $E_m = 200 \times 10^9$ N/m^2, $\rho = 7832$ kg/m^3 and $\nu_p = 0.35$:

$$\Delta p_{max} = 16(200 \times 10^9 \text{N/m}^2)(0.001\text{m})^4 / 9(1-0.35^2)(0.003\text{m})^4 = 5 \times 10^9 \text{ N/m}^2$$
$$= 5000 \text{ MPa}$$

This is a very high pressure limit, but reflects the stiffness of steel and the relatively small radius, thick diaphragm used. A larger radius for a given thickness lowers the natural frequency and the maximum Δp: For example, doubling the radius here lowers Δp by 16 times (r^4) to about 312 MPa.

PROBLEM 9.18

KNOWN: Pressure transmission system filled with air at 20°C
  l = 0.25 m
  d = 3.25 mm
  V = 1600 mm²
  Transducer:
    f_n = 100kHz

FIND: ω_{max} such that $0.9 \leq M(\omega) \leq 1.1$

PROPERTIES: Air: $\mu = 1.8 \times 10^{-5}$ N-s/m²; R = 0.287 kJ/kg-K

ASSUMPTIONS: Air behaves as a perfect gas

SOLUTION

We can use (9.18) as the system model for this process.

$$V_t = \pi d^2 l/4 = 2400 \text{ mm}^3$$

Hence, $V_t > V$. Equations 9.23 and 9.24 are used to estimate the frequency response of the transmission system.

$$\omega_n = \frac{a}{l\left(\frac{1}{2} + 4\frac{V}{V_t}\right)^{1/2}}$$

$$\zeta = \frac{16\mu l}{d^2 \rho a}\left(\frac{1}{2} + 4\frac{V}{V_t}\right)^{1/2}$$

If we assume that the process occurs at a pressure near atmospheric pressure, then we can compute the density of the air by

$$\rho = p/RT = (101325 \text{ N/m}^2 \text{ abs})/(0.287 \text{ kJ/kg-K})(293 \text{ K}) = 1.16 \text{ kg/m}^3$$

The acoustic wave velocity is

$$a = [kRT]^{0.5} = [(1.4)(0.287 \text{ kJ/kg-K})(293 \text{ K})]^{0.5} = 345 \text{ m/s}$$

Then,

$$\omega_n = 775 \text{ rad/s} \qquad \zeta = 0.026$$

and with

$$M(\omega) = 1/[(1 - (\omega/\omega_n)^2)^2 + (2\zeta\omega/\omega_n)^2]^{.5}$$

ω [rad/s]	$M(\omega)$
10	1.00
100	1.02
200	1.07
225	1.09
250	1.12
300	1.17

The frequency response remains within the ±10% constraint over the frequency band, $0 \le \omega \le 230$ rad/s.

COMMENT

Note how the natural frequency of the tubing is much less than that of the transducer. As a consequence, the response characteristics of the connecting tubing govern the system response. The limits of the frequency response of the transducer do not come into play.

The assumption concerning the pressure affects the density of the air only. Its effect on the solution is minimal for low gauge pressures.

PROBLEM 9.20

KNOWN: Pitot-static probe measures flow in a duct.
$H = p_v/\gamma = 20.3$ cm H_2O

FIND: U

ASSUMPTIONS: Duct flow is air at room temperature.

PROPERTIES: $\rho_{air} = 1.16$ kg/m^3
$\rho_{h2o} = 998$ kg/m^3

SOLUTION

$$p_v = p_t - p = 0.5\rho_{air}U^2$$

$$U = [2p_v/\rho_{air}]^{0.5} = [2(\rho_{h2o}g/g_c)H/\rho_{air}]^{0.5}$$

$$= [(2)(998 \text{ kg/m}^3)(9.8 \text{ m/s}^2)(.203 \text{ m } h_2o)/1.16 \text{ kg/m}^3]^{.5} = 58.5 \text{ m/s}$$

PROBLEM 9.21

KNOWN: Pitot-static tube in air.

FIND: U', u_U

SOLUTION

The mean velocity is determined by converting the mean voltage values to pressure and, finally, to velocity. From the given dtaa, the pooled estimates are:

$$<\bar{E}> = (2.438 + 2.354 + 2.473)/3 = 2.422 \text{ V}$$
$$<S_E> = [(.01^2 + 0.009^2 + .012^2)/3]^{.5} = 0.01 \text{ V} \quad \nu_E = 60$$

From the calibration data, p = f(E), so we determine:

$$<\bar{p}> = 0.205 + 0.950E = 2.506 \text{ N/m}^2$$

Now variations in voltage are related to variations in pressure by the static sensitivity at the operating voltage (the mean voltage here), as noted in discussions related to Figures 1.6 and 5.2. Here K_p = dp/dE = 0.95 N/m²/V.

$$<S_p> = K_p<S_E> = 0.0095 \text{ N/m}^2 \quad \text{with } \nu_p = 60$$

The velocity is estimated by $U = (2p_v/\rho)^{.5}$ where p_v is the dynamic pressure measured by the pitot-static tube sensor.

$$<\bar{U}> = [(2)(2.506 \text{ N/m}^2)/1.2 \text{ kg/m}^3]^{.5} = 2.04 \text{ m/s}$$

Variations in pressure are related to variations in velocity. Here K_U = dU/dp = $(2p_v/\rho)^{-.5}$ = 0.41 m/s/N/m² at p_v = 2.506 N/m²:

$$<S_U> = K_U<S_p> = K_U K_p <S_E> = 0.004 \text{ m/s}$$

This value represents the effect of the variations in the test data during repetition. The effect of variation in the measured mean value during replication is estimated by:

$$S_{\bar{U}} = K_U K_p S_{\bar{E}} = K_U K_p [\Sigma(E_j - <E>)^2/2]^{.5} = 0.022 \text{ m/s with } \nu = 2$$

For the voltmeter: (see Tables 5.1 through 5.3 for subscript assignments)

$B_{24} = \pm 10 \ \mu V = \pm (10 \ \mu V)(.95 \ N/m^2/V)(0.41 \ m/s/N/m^2) = 4 \times 10^{-6}$ m/s

For the pitot-static tube, a 1% intrinsic error is assigned:

$B_{22} = 1\% \ K_U P_v = (0.01)(0.41 \ m/s/N/m^2)(2.506 \ N/m^2) = 0.010$ m/s

For the transducer:

B_{23} = negligible (assumed - no info given)

From the calibration data $tS_{yx} = 0.002 \ N/m^2$ with $\nu = 30$, so $S_{yx} \approx .001 \ N/m^2$

$P_{23} = (0.001 \ N/m^2)(.41 \ m/s/N/m^2) = 0.00041$ m/s $\nu_{24} = 30$

For the measurement:

$P_{29} = 0.004$ m/s with $\nu_{29} = 60$ (repetition effect)

$P_{28} = S_U = 0.022$ m/s (replication effect)

Collecting terms:

$B = (B_{22}^2 + B_{23}^2 + B_{24}^2)^{.5} = 0.010$ m/s

$P = (P_{23}^2 + P_{28}^2 + P_{29}^2)^{.5} = 0.024$ m/s with $\nu = 3$ (from 5.25)

so,

$u_U = [B^2 + (tP)^2]^{.5} = 0.07$ m/s

$U' = 2.04 \pm 0.07$ m/s (95%)

PROBLEM 9.22

KNOWN: Pressure measured at M = 4 stations, N = 20 each.

FIND: Effects of data pooling

SOLUTION

Pooling of the data obtained at 4 different radial planes would provide an overall average pressure which accounts for non-axisymmetric effects, that is spatial variations between cross-planes. In fact, by comparing the mean values found along each plane, an estimate of the non-symmetry can be found.

$$<\bar{p}> = (153 + 142 + 161 + 157)/4 = 153.25 \text{ MN/m}^2$$

A measure of the average variation along any cross-section is estimated by

$$<S> = [(7^2 + 9^2 + 9^2 + 7^2)/4]^{.5} = 8.1 \text{ MN/m}^2$$

A measure of the spatial variation in the mean is inferred by

$$S = [\Sigma (\bar{p}_j - <\bar{p}>)^2/3]^{.5} = 8.2 \text{ MN/m}^2 \quad (\text{note: S here is } S_{<\bar{p}>})$$

Note: The closeness of $<S>$ and S here is coincidence).

Replications provide a means to estimate how well the test conditions and their effect on the measured results can be repeated.

PROBLEM 9.23

KNOWN: Air flow measured using a pitot-static tube and a mercury filled manometer.

$10 \leq U \leq 100$ ft/s

FIND: Manometer resolution required to achieve a zero order uncertainty of 5 and 1% in measured velocity.

SOLUTION

The velocity of a flow of fluid of density ρ bringing about a manometer deflection H is given by

$$U = [2(\rho_{Hg}g/g_c)H/\rho]^{0.5} \quad \text{or} \quad H = (0.5\rho U^2)/(\rho_{Hg}g/g_c)$$

At the zero order, our concern is only on the resolution of the measuring instrument. The uncertainty in U depends on the measured H as:

$$(u_0)_U = \pm \left[\left(\frac{\partial H}{\partial u}u_u\right)^2\right]^{1/2} = \pm \left[\left(\frac{\rho u}{\rho_{Hg}g}\right)^2\right]^{1/2}$$

which can be expressed as,

$$(u_0)_U/U = \pm u_H/2H$$

To achieve $u_U/U = 0.05$ requires: $(u_0)_H/H = \pm 0.10$, or an interpolation error of 10% of the expected manometer deflection. Assuming that u_0 is 1/2 the resolution, a resolution of 20% of the expected manometer deflection is needed.

U [m/s]	H [cm Hg]	Resolution [cm]
5	0.011	0.002
50	1.128	0.226

Using $u_U/U = 0.01$, $(u_0)_H/H = \pm 0.02$, or a resolution of 4% of the expected manometer deflection.

PROBLEM 9.24

KNOWN: Static pressure around cylinder, $p(\theta)$.
$p_v = 8$ in H_2O
$p_{atm} = 14.7$ psia
$T_{atm} = 60°F$

FIND: $U(\theta)$

ASSUMPTIONS: Total pressure constant throughout tunnel.

PROPERTIES: Water: $\rho = 62.4$ lb_m/ft^3
Air: $\rho = p/RT = 0.075$ lb_m/ft^3

SOLUTION

The Bernoulli equation can be written along a streamline from the freestream to the stagnation point as

$$p_t = p_\infty + 0.5\rho U^2_\infty$$

Similarly,

$$p_\infty + 0.5\rho U^2_\infty = p(\theta) + 0.5\rho U^2(\theta) \text{ or}$$
$$p_t = p(\theta) + 0.5\rho U^2(\theta) \text{ Then,}$$

$$U(\theta) = [2(p_t - p(\theta))/\rho_{air}]^{.5}$$

The manometer will provide a deflection relative to the pressure difference, $p(\theta) - p_{atm}$. But at the $0°$ tap, $p(\theta) = p_t$, so the manometer will respond to the pressure difference: $p_t - p_{atm}$, or

$$H(\theta = 0°) = (p_t - p_{atm})/\gamma_{h2o}$$

But $H(0) = 0$, so that $p_t = p_{atm}$. This allows us to rewrite the expression for velocity as (where H [inch] and U [ft/s]),

$$U(\theta) = [0.5 (p_{atm} - p(\theta))/\rho]^{0.5} = U(\theta) = [0.5\rho_{h2o}gH/\rho_{air}]^{.5} = 66.8H^{.5}$$

θ	H [in H_2O]	U [ft/s]
0	0	0

θ	H (in H₂O)	U (ft/s)
45	16.3	270
90	32.0	378
135	9.1	201
180	9.4	205

PROBLEM 9.25

KNOWN: Pitot-static probe. $T_{amb} = 20\ °C$

FIND: Lowest airspeed and manometer deflection for which the viscous correction is negligible.

PROPERTIES: Air: $\rho = 1.225\ kg/m^3$
$\nu = 2 \times 10^{-5}\ m^2/s$
Water: $\gamma_m = 9780\ N/m^2$

SOLUTION

Viscous correction is required when $Re_r < 500$. For $Re_r > 500$,

$$Re_r = Ur/\nu\ \text{So,}$$
$$U > 500\nu/r = (500)(2 \times 10^{-5}\ m^2/s)/r\ \text{or}$$
$$U > (0.01/r)\ m/s$$

From the pitot-static probe,

$$U = [2\gamma_m H/\rho]^{0.5}$$

Equating with the above expression for velocity (with r and H in [m]) gives,

$$H > (\rho/2\gamma_m)(500\nu/r)^2 = ([1.225\ kg/m^3]/[2][9780\ N/m^3])(0.01/r)^2 = 6.26 \times 10^{-9}/r^2$$

For example, for a 6 mm diameter probe, $U > 3.3\ m/s$ (10.8 ft/s) and $H > 0.7\ mm$.

PROBLEM 9.26

KNOWN: Anemometer circuit in Figures 9.25 and 9.26
$R_3 = 500 \, \Omega$; $R_4 = 500 \, \Omega$; $\alpha = 0.00395/^\circ C$
Sensor resistance at temperature T_s: $R_s(20^\circ C) = 110 \, \Omega$

FIND: R_D if $T_s = 60^\circ C$

SOLUTION

For a metallic resistance temperature device, the resistance temperature relation is approximated by,

$$R_s(T_s) = R_o[1 + \alpha(T_s - T_o)]$$

Using $R_o = R(20^\circ C) = 110 \, \Omega$

$$R_s(60^\circ C) = 110\Omega[1 + 0.00395(60 - 20)]$$

$$= 127.38 \, \Omega$$

If the sensor resistance is R_s, the bridge will be balanced when

$$R_s = R_D(R_3/R_4) \quad \text{so that} \quad R_D = 127.38 \, \Omega.$$

PROBLEM 9.27

KNOWN: Constant resistance anemometer

FIND: Sensitivity relative to velocity

SOLUTION

For a thermal anemometer operating at constant resistance,

$$E^2 = A + BU^n$$

The sensitivity, K, is found by

$$K = (dE/dU)_U = nBU^{n-1}/2(A+BU^n)^{0.5}$$

For n = 0.5,

$$K \propto U^{-0.75}$$

The sensitivity decreases as velocity increases.

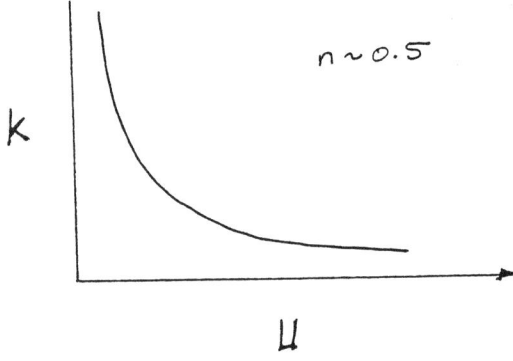

PROBLEM 9.28

KNOWN: F = 600 mm
θ = 5.5º
λ = 514.5 nm

FIND: f_d at U = 1, 10, 100 m/s

SOLUTION

$$f_d = U [2 \sin \theta/2]/\lambda = U [2 \sin 2.75º]/514.5 \times 10^{-9} m = 186{,}540 \, U$$

At U = 1 m/s, f_d = 186.540 kHz
U = 10 m/s, f_d = 1.8654 MHz
U = 100 m/s, f_d = 18.654 MHz

Recomputing for F = 300 mm and θ = 7.3º,

$$f_d = 247{,}517 \, U$$

At U = 1 m/s, f_d = 247.517 kHz
U = 10 m/s, f_d = 2.47517 MHz
U = 100 m/s, f_d = 24.7517 MHz

PROBLEM 9.29

KNOWN: LDA measurement using dual beam mode.
 N = 5000
 U = 21.37 m/s
 S_U = 0.43 m/s (see Note)
 λ = 623.8 ± 0.5% nm
 $B_{fd}/f_d \leq 0.9\%$
 $B_\theta \leq 0.25°$

FIND: Estimate U', the best estimate of the velocity, based on available information.

ASSUMPTIONS: To do this problem, we need a value for θ. Let θ = 6°.

SOLUTION

The relation between velocity and Doppler frequency using dual beam mode is,

$$U = f_d \lambda / [2 \sin \theta/2]$$

with the mean Doppler frequency estimated by

$$f_d = U[2 \sin \theta/2]/\lambda = (21.37 \text{ m/s})[2 \sin 3°]/(623.8 \times 10^{-9} \text{ m})$$

$$= 3.5858 \text{ MHz}$$

The bias error in the velocity estimate can be estimated by

$$B_U = \pm \left[\left(\frac{\partial U}{\partial f_d} B_{f_d}\right)^2 + \left(\frac{\partial U}{\partial \lambda} B_\lambda\right)^2 + \left(\frac{\partial U}{\partial \theta} B_\theta\right)^2 \right]^{1/2}$$

Using

 B_θ = 0.25° = 0.0044 rad
 B_{fd} = 0.009 f_d = 32,272 Hz
 B_λ = (623.8 × 10⁻⁹ m)(0.005) = 3.12 × 10⁻⁹ m

$$B_U = \pm\left[\left(\frac{\lambda}{2\sin\frac{\theta}{2}}B_{f_d}\right)^2 + \left(\frac{f_d}{2\sin\frac{\theta}{2}}B_\lambda\right)^2 + \left(\frac{f_d \lambda}{\sin\frac{\theta}{2}\tan\frac{\theta}{2}}B_\theta\right)^2\right]^{1/2}$$

$$= \pm\left[\left(\frac{623\times10^{-9}\,m}{2\sin 3°}9836/s\right)^2 + \left(\frac{1.09\times10^6/s}{2\sin 3°}3.12\times10^{-9}\,m\right)^2 + \left(\frac{1.09\times10^6/s}{\sin 3°}\frac{623.8\times10^{-9}\,m}{\tan 3°}\times.0044\right)^2\right]^{1/2}$$

$$= [0.19^2 + 0.11^2 + 3.27^2]^{0.5} = \pm 3.28 \text{ m/s}$$

The bias error is completely dominated by the optical system error.
The precision error in the mean velocity is estimated by

$$P_U = S_U/N^{0.5} = (0.43 \text{ m/s})/5000^{0.5} = 0.007 \text{ m/s}$$

with degrees of freedom of 4999.

The uncertainty in velocity is estimated by

$$u_U = \pm[B^2 + (t_{4999,95}P_U)^2]^{0.5} \quad (95\%)$$

$$= \pm[3.28^2 + (1.96 \times .007)^2]^{0.5} = 3.28 \text{ m/s} \quad (95\%)$$

The best estimate of the velocity may be stated as

$$U' = 21.37 \pm 3.28 \text{ m/s} \quad (95\%)$$

NOTE: The first printing gave $S_U = 0.43$ ft/s. This makes $P_U = 0.002$ m/s but its affect on the above result is negligible.

COMMENT

The error in the optical set-up dominates the overall uncertainty in measuring velocity. One can minimize this error by using a shorter focal length lens, as this will increase the value of θ. But the value assigned for the bias error in θ in this problem is actually quite large for precision work. Standardized metrology equipment and methods allow vendors of high quality optics to measure θ to better than 0.01°. At $B_\theta = 0.01°$,
$U' = 21.37 \pm 0.26$ m/s (95%), a 1% uncertainty.

PROBLEM 10.1

KNOWN: Flow of air through a pipe
$U(r) = 25[1 - (r/r_1)^2]$ cm/s
$p = 14.7$ psia $= 101{,}300$ N/m² abs
$T = 5°C = 278$ K
$d_1 = 2r_1 = 5$ cm

FIND: mass flow rate

ASSUMPTIONS: Steady, incompressible, axisymmetric flow of a perfect gas.

SOLUTION

Conservation of mass gives

$$\frac{\partial}{\partial t}\iiint_{CV} \rho\, d\forall + \iint_{CS} \rho\, \vec{U} \cdot \hat{n}\, dA = 0$$

which for steady, incompressible, axisymmetric flow becomes

$$\dot{m} = \int_0^{2\pi}\int_0^{r_1} \rho\, U(r)\, r\, dr\, d\theta$$

The velocity can be written in vector form as

$$\vec{U} = 25[1 - (r/r_1)^2]\, \hat{e}_z$$

so that

$$\dot{m} = \rho \int_0^{2\pi}\int_0^{r_1} \left(25\left(1 - (r/r_1)^2\right)\right) r\, dr\, d\theta$$

$$= \frac{25}{2}\rho\pi r_1^2 = 39.3\rho \text{ cm}^3/\text{s}$$

For a perfect gas, the density can be estimated by

$$\rho = p/RT = 1.16 \text{ kg/m}^3$$

so that (with 1 m = 100 cm):

$$\dot{m} = 4.5 \times 10^{-5} \text{ kg/s} = 0.16 \text{ kg/h}$$

PROBLEM 10.2

KNOWN: Air flow through a pipe
$2r_1 = 20$ cm
N = 5 velocity measurements per cross-sectional traverse
M = 3 cross-sectional traverses

FIND: Q'

ASSUMPTIONS: Steady, incompressible flow of a perfect gas.
Flow rate remains perfectly controlled (constant) during all measurements (equivalent to the assumption that all measurements are taken simultaneously).
Bias errors are negligible.

SOLUTION

The flow rate along any traverse line can be approximated by

$$Q_j = 2\pi \Sigma_i U_{ij} r \Delta r \quad i = 1,2,3,4,5$$

where $\Delta r = 2$ cm and $j = 1, 2, 3$. Then,

$$Q_1 = 2\pi[(25.31)(1)(2) + (22.48)(3)(2) + (21.66)(5)(2) + (15.24)(7)(2) + (5.12)(9)(2)]$$

$$= 4446 \text{ cm}^3/\text{s}$$

Similarly, $Q_2 = 4421$ cm³/s, $Q_3 = 4400$ cm³/s. The mean flow rate is

$$\overline{Q} = (1/3)[4446 + 4421 + 4400] \text{cm}^3/\text{s} = 4423 \text{ cm}^3/\text{s}$$

with standard deviation,

$$S_Q = \left[\frac{1}{2} \sum_{j=1}^{3} (Q_j - \overline{Q})^2 \right]^{1/2}$$

$$= 23 \text{ cm}^3/\text{s}$$

and

$$S_{\overline{Q}} = S_Q/3^{.5} = 13.3 \text{ cm}^3/\text{s} \quad \text{with } \nu = 2$$

Then, $t_{2,95} = 4.303$ and

$$Q' = 4423 \pm 57.2 \text{ cm}^3/\text{s} \quad (95\%)$$

COMMENT

The assumption of negligible bias error would most likely not be realistic in practice. Examples of possible bias error in such a measurement include: instrument errors, maintenance of a perfectly controlled flow rate during all measurements, measuring probe placement errors and data reduction error (the approximation of equation 10.5).

PROBLEM 10.3

KNOWN: Manometer measuring the pressure drop of flowing water
$H = 10.16$ cm Hg
$d_1 = 5.1$ cm
$\gamma = 9800$ N/m^3 (water)
$S_m = 13.57$ (mercury) or $\gamma_m = \gamma S_m$

FIND: $p_1 - p_2$

ASSUMPTIONS: p_1 and p_2 taps are located along the same horizontal datum line.

SOLUTION

Applying the hydrostatic equation between points 1 and 2 yields

$p_1 + L\gamma + H\gamma_m - (L+H)\gamma = p_2$

$p_1 - p_2 = H(\gamma_m - \gamma)$

$= H(13.57\gamma - \gamma) = H\gamma 12.57$

$= (0.1016 \text{ m})(12.57)(9800 \text{ N/m}^2)(1 \text{ Pa/N/m}^2) = 12.516$ kPa

PROBLEM 10.4

KNOWN: Air flow through orifice meter
$p_1 - p_2 = 69$ kPa
$d_1 = 25.4$ cm
$T = 32$ °C

FIND: H

SOLUTION

$$p_1 - p_2 = \gamma H$$

$$H = (p_1 - p_2)/\gamma$$

with $\gamma = 9750$ N/m^3 (Appendix C)

$H = (69{,}000 \text{ N/m}^2)/9750 \text{ N/m}^3 = 7.041$ m H$_2$O $= 704.1$ cm H$_2$O

PROBLEM 10.5

KNOWN: Water flow through orifice meter using flange taps.

$d_1 = 3$ in.
$T = 60°F$
$p_1 = 100$ psi
$p_2 = 76$ psi
$d_o = 1.5$ in
$R_{O2} = 48.3$ ft-lb/lb$_m$-°R

FIND: Is $Y < 1$?

ASSUMPTIONS: Steady flow

SOLUTION

$$Y = f(k, \beta, \Delta p/p_1)$$

For O_2 (and any diatomic gas), $k = 1.4$.

$$\beta = d_o/d_1 = 0.5$$

$$\Delta p/p_1 = 0.24$$

Using Figure 10.7,

$$Y = 0.92$$

There is an 8% reduction in flow rate due to compressibility effects.

COMMENT

Aside from the usual sources of error in an measurement, data reduction errors enter into the obstruction meter relations from the assumed values of the various coefficients and from the ability to read these values from the tables and charts.

PROBLEM 10.6

KNOWN: Orifice meter using flange taps
$d_o = 5$ cm
$d_1 = 15$ cm
$Re_{d1} = 250{,}000$

FIND: C

SOLUTION

$$C = f(\beta, Re_{d1})$$

$$\beta = d_o/d_1 = 0.33$$

$$Re_{d1} = 250{,}000$$

From Figure 10.6, $K_o = CE = 0.60$.

$$E = 1/(1 - \beta^4)^{0.5} = 1.006$$

$$C = K_o/E = 0.596 \approx 0.60$$

COMMENT

Aside from the usual sources of error in an measurement, data reduction errors enter into the obstruction meter relations from the assumed values of the various coefficients and from the ability to read these values from the tables and charts.

PROBLEM 10.7

KNOWN: Flow of water through orifice meter at 20°C
$d_1 = 10$ cm
$\beta = 0.4$

FIND: Q at which $C = f(\beta, Re_{d1})$ becomes $C = f(\beta)$

ASSUMPTIONS: Flange pressure taps are used so that Figure 10.6 is applicable.

SOLUTION

For $\beta = 0.4$, Figure 10.6 indicates a Reynolds number independence in flow coefficient for $Re_{d1} > 20{,}000$. Because $K_o = CE$ and E depends only on β, we conclude that $C = f(\beta)$ for all $Re_{d1} > 20{,}000$.

$$Re_{d1} = 4Q/\pi d_1 \nu > 20{,}000$$

$$Q > \pi \nu d_1 Re_{d1}/4 = \pi(1 \times 10^{-6} \text{ m}^2/\text{s})(0.1\text{m})(20000)/4 = 1.6 \times 10^{-4} \text{ m}^3/\text{s}$$

with ν from Appendix C.

COMMENT

Aside from the usual sources of error in an measurement, data reduction errors enter into the obstruction meter relations from the assumed values of the various coefficients and from the ability to read these values from the tables and charts.

PROBLEM 10.8

KNOWN: Water flow through an orifice plate with flange taps
$Q = 50$ L/s
$T = 25°C$
$d_1 = 12$ cm
$\beta = 0.5$

FIND: Δp, $(\Delta p)_{loss}$

ASSUMPTIONS: Steady, incompressible ($Y = 1$) flow.

SOLUTION

From (10.14), $Q = CEAY(2\Delta p/\rho_1)^{.5}$ with A and β based on d_o. Now, $d_o = \beta d_1 = 0.06$m and $A = \pi d_o^2/4 = .0028$m^2:

$$K_o = CE = f(\beta, Re_{d1}) = f(0.5, 530{,}000) \approx 0.63$$

with $\nu = 1 \times 10^{-6}$ m^2/s and $\rho = 997$ kg/m^3 from Appendix C,

$$Re_{d1} = 4Q/\pi\nu d_1 = 4(50 \times 10^{-3}\text{m}^3)/\pi(1 \times 10^{-6}\text{m}^2/\text{s})(0.12\text{m}) = 530000$$

Then,

$$\Delta p = (\rho_1/2)(Q/CEAY)^2$$
$$= [(1000 \text{ kg/m}^3)/2]\{(0.05\text{m}^3/\text{s})/(0.63)(0.0028\text{m}^2)(1)\}^2$$
$$= 4.017 \times 10^5 \text{ N/m}^2$$

From Figure 10.8, we can expect

$$(\Delta p)_{loss} = 0.77 \Delta p = 3.093 \times 10^5 \text{ N/m}^2$$

COMMENT

Aside from the usual sources of error in an measurement, data reduction errors enter into the obstruction meter relations from the assumed values of the various coefficients and from the ability to read these values from the tables and charts.

PROBLEM 10.9

KNOWN: Air flow through a flow nozzle (k = 1.4)
d_o = 3 cm
d_1 = 6 cm
H = 75 cm H_2O
p_1 = 94.4 kPa
T_1 = 38 °C

FIND: Q

ASSUMPTIONS: Steady flow of a perfect gas.
ASME long-radius nozzle with throat taps (so that Fig. 10.7 and 10.11 are applicable)

SOLUTION

From (10.14), $Q = CEAY(2\Delta p/\rho_1)^{.5}$ with β and A based on the throat diameter d_o. Then, $\beta = d_o/d_1 = 0.5$ and $A = \pi d_o^2/4 = 0.00071$ m². With γ = 9800 N/m³ from Appendix C,

$$\Delta p = \gamma H = (9800 \text{ N/m}^3)(0.75 \text{ m}) = 7350 \text{ Pa}$$

$$\Delta p/p_1 = 7350 \text{ Pa}/94900 \text{ Pa} = 0.078$$

From Figure 10.7, $Y = f(k,\beta,\Delta p/p_1) \approx 0.98$. For a perfect gas,

$$\rho_1 = p_1/RT_1$$
$$= (94,400 \text{ Pa})/(287 \text{ J/kgK})(311) = 1.06 \text{ kg/m}^3$$

We need a value for K_o. But $K_o = f(\beta, Re_{d1})$ and Re_{d1} depends on Q. Guess a value for K_o and iterate on a solution. From Figure 10.11, pick K_o = 1.008 from the flat region of the β = 0.5 curve. Then,

$$Q = (1.008)(0.00071 \text{m}^2)(0.98)[(2)(7350 \text{ Pa})/(1.06 \text{kg/m}^3)]^{.5} = 0.082 \text{ m}^3/\text{s}$$

Now check on the guessed value of K_o,

$$Re_{d1} = 4Q/\pi \nu d_1 = (4)(0.078 \text{ cms})/(\pi)(0.06\text{m})(1.6 \times 10^{-5} \text{ m}^2/\text{s}) = 103,000$$

or from Figure 10.11, $K \approx 1.008$. So the solution remains: $Q = 0.082 \text{ m}^3/\text{s}$.

COMMENT

Aside from the usual sources of error in an measurement, data reduction errors enter into the obstruction meter relations from the assumed values of the various coefficients and from the ability to read these values from the tables and charts.

PROBLEM 10.10

KNOWN: Nitrogen (k = 1.4) flows through an orifice meter with flange taps.

$T_1 = 520°R$
$p_1 = 20$ psia
$p_2 = 15$ psia
$d_1 = 4$ in.
$\beta = 0.5$
$R_{N2} = 55.13$ ft-lb/lb$_m$-°R

FIND: Q

ASSUMPTIONS: Steady flow of a perfect gas

SOLUTION

From (10.14): $Q = CEAY(2\Delta p/\rho_1)^{.5}$ where β and A based on the orifice diameter, d_o. Then, $d_o = \beta d_1 = 2$ in. and $A = 0.0218$ ft^2.

$p_1 - p_2 = 20 - 15$ psia $= 5$ psia $= 720$ psf

$(p_1 - p_2)/p_1 = 0.25$

$Y(k, (p_1 - p_2)/p_1, \beta) = Y(1.4, 0.25, 0.5) = 0.92$ (Fig. 10.7)

$\rho_1 = p_1/RT_1$
$= (20 \text{ psi})(144 \text{ in}^2/\text{ft}^2)/(55.13 \text{ ft-lb/lb}_m\text{-°R})(520°R) = 0.1 \text{ lb}_m/\text{ft}^3$

Now, $K_o = CE = f(\beta, Re_{d1})$ and Re_{d1} depends on Q. Guess $K_o = 0.63$ (flat region of Figure 10.6 at $\beta = 0.5$),

$Q = (0.63)(0.0218 \text{ ft}^2)(0.92)$
$\quad \times [(2)(32.2 \text{ lb}_m\text{-ft/lb-s}^2)(5 \text{ lb/in}^2)(144 \text{ in}^2/\text{ft}^2)/0.1 \text{lb}_m/\text{ft}^3]^{.5}$
$= 8.6$ cfs

Checking: $Re_{d1} = 4(8.6 \text{ ft}^3/\text{s})/\pi(1.15 \times 10^{-4} \text{ft}^2/\text{s})(4/12 \text{ ft}) = 285655$
and $K_o \approx 0.63$.

So, $Q = 8.6$ ft^3/s.

PROBLEM 10.11

KNOWN: Water flow through an orifice meter
  T = 20°C
  d_1 = 38 cm
  $\dot{m} \approx$ 200 kg/s
  H ≤ 15 cm Hg (S = 13.57)

FIND: Design orifice meter by setting d_o.

ASSUMPTIONS: Steady, incompressible (Y = 1) flow

SOLUTION

From (10.3), (10.6) and (10.14), the mass flow rate for a constant density flow

$$\dot{m} = \rho_1 Q = \rho_1 CEAY(2\Delta p/\rho_1)^{.5}$$

with A and β based on orifice diameter, d_o. Solving for d_o with A = $\pi d_o^2/4$,

$$d_o = [4\dot{m}/\{CE\pi\rho_1(2\Delta p/\rho_1)^{.5}\}]^{.5}$$

From Appendix C, ρ_1 = 999 kg/m³ and μ = 9.6 x 10⁻⁴ N-s/m².

$$\Delta p = \gamma_{hg}H = S\rho gH \le (13.57)(9760 \text{ N/m}^3)(0.15\text{m}) = 19928 \text{ N/m}^2$$

If we select β = 0.5, d_o = 19 cm. Then

$$Re_{d1} = 4\dot{m}/\pi\mu d_1 = 4(200 \text{ kg/s})/\pi(0.38\text{m})(9.6\text{x}10^{-4}\text{N-s/m}^2) = 698000$$

From Figure 10.6, K_o = f(β, Re_{d1}) = f(0.5, 698000) $\approx$ 0.625, then,

$$d_o \ge [4(200\text{kg/s})/\{(.625)\pi(1000\text{kg/m}^3)[2(19928\text{N/m}^2/(10000\text{kg/m}^3)]^{.5}\}]^{.5}$$

$$= 0.25 \text{ m or } 25 \text{ cm}$$

or $\beta \ge$ 0.658. Using β = 0.658, d_o = 25 cm., K_o(0.658, 698000) $\approx$ 0.625. Hence, this value for d_o will meet our constraints. We may wish to build in a safety factor on the Δp constraint. So setting β = 0.7 or d_o = 26.6 cm (Note that from Figure 10.6 that this is the largest value for β at which the ASME tables may be used. Larger β values will require in situ calibration.), we find K_o = 0.668 and the resulting pressure head works out to 10.8 cm Hg at 200 kg/s.

COMMENT

It should become apparent that by choosing different values for d_o, the pressure drop across an obstruction meter can be altered as desired for an expected flow rate.

PROBLEM 10.12

KNOWN: Water flowing through a venturi meter
$T = 60°F$
$d_1 = 4$ in.
$H \leq 30$ in. H_2O
$Q \approx 17$ cfm $= 0.283$ cfs

FIND: Design the meter size, d_o

ASSUMPTIONS: Steady, incompressible ($Y = 1$) flow.
Cast venturi meter.

SOLUTION

From (10.14), $Q = CEAY(2\Delta p/\rho_1)^{.5}$ where A and β are based on the venturi throat diameter, d_o. Using $\rho = 62.3$ lb_m/ft^3 and $\nu = 1.2 \times 10^{-5}$ ft^2/s for water from Appendix C:

$$\Delta p = \gamma_{H2O}H = (\rho g/g_c)H \leq (62.3 \text{ lb/ft}^3)(30/12 \text{ ft}) = 156 \text{ lb/ft}^2$$

We seek,

$$d_o \geq [4Q/\{CE\pi(2\Delta p/\rho_1)^{.5}\}]^{.5}$$

Using $Re_{d1} = 4Q/\pi\nu d_1 = 4(0.283\text{cfs})/\pi(4/12\text{ft})(1.2 \times 10^{-5}\text{ft}^2/\text{s}) = 265000$, we will choose $\beta = 0.5$ such that $C = 0.984$, $E = 1/(1 - \beta^4)^{0.5} = 1.0328$. Then,

$$d_o \geq [4(0.283\text{cfs})/\{(.984)(1.0328)\pi$$

$$\times [2(156\text{psf})(32.2\text{lb}_m\text{-ft/lb-s}^2)/62.3\text{lb/ft}^3)^{.5}\}]^{.5} = 0.167 \text{ ft}$$

This yields a β sufficiently close to 0.5, as selected. A value of $d_o \geq 2.1$ in. will meet the constraint.

COMMENT

It should be apparent that by choosing different values for d_o, the pressure drop across an obstruction meter can be altered, as desired, for an expected flow rate to achieve a desired range of values.

PROBLEM 10.13

KNOWN: Q = 120 cfm of water at 60°F
H ≤ 20 in. Hg
d_1 = 6 in.
pump efficiency: η = 0.60
power costs: $ 0.10/kW-hr
operating time: 6000 hr/yr

FIND: Specify design for suitable orifice, venturi and nozzle.
$(\Delta p)_{loss}$ and operating costs for each device.

ASSUMPTIONS Steady, incompressible (Y = 1) flow.

PROPERTIES water: ρ = 62.4 lb_m/ft^3 ν = 1 x 10^{-5} ft^2/s

SOLUTION

From equation 10.14, $Q = CEAY(2\Delta p/\rho_1)^{.5}$ where $\Delta p = \rho g H/g_c$ = 1411 lb/ft^2. At Q = 120 cfm,

$$Re_{d1} = 4Q/\pi d_1 \nu = 4(120 cfm)(1 min/60s)/\pi(.5 ft)(1 \times 10^{-5} ft^2/s) = 5.1 \times 10^5$$

The meter throat sizes d_o are found from

$$d_o = [4Q/\{CE\pi(2\Delta p/\rho_1)^{.5}\}]^{.5}$$

Orifice:

From Figure 10.6, $K_o = f(\beta, Re_{d1})$. If we guess, K_o = 0.6, then d_o = 0.33 ft for $\beta = d_o/d_1$ = 0.66. From Figure 10.6, $K_o(0.66, 5 \times 10^5)$ = 0.65, so that d_o = 0.32 ft for β = 0.64. Then from Figure 9.6, $K_o(0.64, 5 \times 10^5)$ = 0.65. So β = 0.64 and
$$d_o = 3.84 \text{ in.}$$

From Figure 10.8, with β = 0.64 and $Re_{d1} = 5 \times 10^5$,

$$(\Delta p)_{loss} = 0.6 \Delta p = 847 \text{ psf}$$

ASME Long Radius Nozzle:

From Figure 10.11, $K_o = f(\beta, Re_{d1})$. If we guess, $K_o = 1.0$, then $d_o = 0.26$ ft for $\beta = d_o/d_1 = 0.52$. From Figure 10.11, $K_o(0.52, 5 \times 10^5) = 1.01$, so that $d_o = 0.26$ ft. So $\beta = 0.52$ and

$$d_o = 3.12 \text{ in.}$$

From Figure 10.8, with $\beta = 0.52$ and $Re_{d1} = 5 \times 10^5$,

$$(\Delta p)_{loss} = 0.6 \, \Delta p = 847 \text{ psf} \quad \text{(Same as the orifice, but with smaller } \beta\text{)}$$

Venturi:

Suppose we choose a 15° cast model. The text states, based on the ASME Power Test Codes, that $C \approx 0.98$. So we begin with a guess, $K_o = CE = 0.98$. Then, $d_o = 0.265$ for a $\beta = 0.53$. $E = (1-\beta^4)^{-1/2} = 1.04$, so $K_o = 1.02$. Using $K_o = 1.02$, $d_o = 0.26$ ft, $\beta = 0.52$ and $E = 1.04$. This is close enough. So,

$$d_o = 3.12 \text{ in.}$$

From Figure 10.8,

$$(\Delta p)_{loss} = 0.18 \, \Delta p = 254 \text{ psf}$$

Cost Analysis

The power use is estimated from:

$$W = Q(\Delta p)_{loss}/\eta$$

and operating cost from:

$$\text{cost} = (W)(\text{operating time})(\text{cost/kW-unit time})$$

	β	W[kW]	Cost[$/yr]
orifice	0.64	3.7	2230
nozzle	0.52	3.7	2230
venturi	0.52	0.9	538

PROBLEM 10.14

KNOWN: Water flow at 80°F through an ASME long radius nozzle.
 $\beta = 0.6$
 $H = 10$ in Hg
 $d_1 = 6$ in.

FIND: Q

ASSUMPTIONS: Steady, incompressible (Y=1) flow.

PROPERTIES: Water: $\rho = 62.2$ lb$_m$/ft^3 $\nu = 1 \times 10^{-5}$ ft^2/s

SOLUTION

From (10.14) with A and β based on throat diameter d_o,

$$Q = CEAY(2\Delta p/\rho_1)^{.5}$$

With $\beta = d_o/d_1 = 0.6$, $d_o = (0.6)(6)$ in. $= 3.6$ in. And $\Delta p = \rho g H/g_c = 703$ psf. The value for $K_o = f(\beta, Re_{d_1})$. From Figure 10.11, guess $K_o = 1.03$. Then, with $A = \pi(3.6/12)^2/4 = 0.071$ ft^2:

$Q = (1.03)(0.071$ ft$^2)[2(703$ psf$)(32.2$ lb$_m$-ft/lb-s$^2)/62.2$ lb$_m$/ft$^3)]^{0.5}$
 $= 1.96$ cfs

Then, $Re_{d_1} = 4(1.96$ cfs$)/\pi(.5$ ft$)(1 \times 10^{-5}$ ft^2/s$) = 5 \times 10^{-5}$

From Figure 10.11, $K_o = (0.6, 5 \times 10^{-5}) = 1.03$. So $Q = 1.96$ ft^3/s.

PROBLEM 10.15

KNOWN: Orifice meter
$d_1 = 9$ cm
$d_o = 4$ cm
Meter to be located within 4 m of straight run of pipe.
Upstream elbow.

FIND: Meter placement

ASSUMPTIONS: Elbow is in-plane with downstream pipe.

SOLUTION

For $\beta = d_o/d_1 = 0.44$, we use Figure 10.13 which indicates that the meter should be located in a straight run of pipe 6.25 diameters downstream of the elbow with 3 diameters of straight run pipe downstream of the meter. Using 7 diameters upstream to be conservative, this should just fit into the available space.

PROBLEM 10.16

KNOWN: Orifice meter used as a sonic nozzle to meter air flow.
$T_1 = 100°F$
$p_1 = 13.4$ psia
$p_2 = 6.4$ psia
$d_1 = 2$ in.
$\beta = 0.4$

FIND: $\dot{m}$

PROPERTIES: $R = 53.3$ ft-lb/lb$_m$-°R $k = 1.4$

ASSUMPTIONS: Air behaves as a perfect gas.

SOLUTION

For this flow,

$$p_2/p_1 = 6.4\text{psia}/13.4\text{psia} = 0.477$$

$$(p_2/p_1)_{crit} = [2/(k+1)]^{k/(k-1)} = 0.528$$

Since $p_2/p_1 < 0.528$, the flow in the orifice throat is choked and sonic conditions exist there. Using,

$$\rho_1 = p_1/RT_1 = (13.4\text{psi})(144\text{ in}^2/\text{ft}^2)/(53.3\text{ ft-lb/lb}_m\text{-°R})(560°R)$$

$$= 0.065 \text{ lb}_m/\text{ft}^3$$

$$A = \pi d_o^2/4 = 0.0035 \text{ ft}^2$$

Equation 10.18 applies using $k = 1.4$, $R = 53.3$ ft-lb/lb$_m$-°R, and $T = 560°R$:

$$\dot{m} = \rho_1 A (2RT_1)^{.5}[(k/k+1)(2/k+1)^{2/k+1}]^{.5} = 0.15 \text{ lb}_m/\text{s}$$

PROBLEM 10.17

KNOWN: Air flow through an orifice meter with flange taps.
$d_o = 0.5$ m
$d_1 = 1$ m
$H = 90$ mm H_2O
$p_1 = 2$ atm $= 3$ atm absolute

FIND: Q

ASSUMPTIONS: Steady flow of a perfect gas.

PROPERTIES: air: $\nu = 1.5 \times 10^{-4}$ m^2/s $R = 0.287$ kJ/kg-K
water: $\rho = 1000$ kg/m^3

SOLUTION

Using equation 10.14 with A and β based on the orifice throat diameter d_o,

$$Q = CEAY(2\Delta p/\rho_1)^{.5}$$

where
$\rho_1 = p_1/RT_1 = (3)(101{,}325$ N-m^2/atm$)/(287$ J/kg-K$)(293$K$) = 3.6$ kg/m^3

$\Delta p = \rho gH/g_c = (1000$ kg/m$^3)(9.8$ m/s$^2)(0.09$ m$)/1$ kg-m/N-s$^2 = 882$ N/m^2

With $\beta = d_o/d_1 = 0.5$ and $\Delta p/p = 0.003$, Figure 10.7 gives $Y \approx 1$. From Figure 10.6, guess $K_o = f(\beta, Re_{d1}) = 0.633$. Then,

$$Q = (0.633)(\pi (0.5 \text{ m})^2/4)[2(887 \text{ N/m}^2)/1000 \text{ kg/m}^3]^{.5} = 2.75 \text{ m}^3/\text{s}$$

Checking, $Re_{d1} = 4Q/\pi d_1 \nu = 4(2.75$ m^3/s$)/\pi(1$m$)(1.5 \times 10^{-4}$m^2/s$) = 23{,}350$

From Figure 10.6, $K_o(0.5, 23350) \approx 0.635$ or $Q = 2.76$ m^3/s. Repeating, $Re_{d1} = 23{,}400$ and $K_o = 0.635$. So, $Q = 2.76$ m^3/s.

PROBLEM 10.18

KNOWN: Flow of 70°F water through an ASME Long Radius Nozzle.
 $\beta = 0.5$
 $d_1 = 3$ in.
 $10 \le Q \le 25$ gpm $(0.022 \le Q \le 0.056$ ft^3/s)
 Transducer:
  $(u_d)_{\Delta p}/\Delta p$: 0.25% full scale

FIND: Δp or H; design stage uncertainty in Q

ASSUMPTIONS: Steady, incompressible (Y = 1) flow of water.

PROPERTIES: Water: $\rho = 62.4$ lb$_m$/ft^3 $\nu = 1.2 \times 10^{-5}$ ft^2/s

SOLUTION

Rewriting equation 10.14,

$$\Delta p = (\rho_1/2)(Q/CEYA)^2$$

Using $d_o = \beta d_1 = 1.5$ in. and $A = \pi d_o^2/4 = \pi(1.5/12)/4 = 0.0123$ ft^2:

At 10 gpm, $Re_{d1} = 4Q/\pi d_1 \nu = 4(0.022 \text{ cfs})/\pi(.25 \text{ ft})(1.2 \times 10^{-5} \text{ ft}^2/\text{s}) = 9500$

$$K_o(0.5, 9500) = 0.965$$

Then, $\Delta p = 3.43$ psf

At 25 gpm, $Re_{d1} = 4Q/\pi d_1 \nu = 4(0.056 \text{ cfs})/\pi(.25 \text{ ft})(1.2 \times 10^{-5} \text{ ft}^2/\text{s}) = 23600$

$$K_o(0.5, 23600) = 0.980$$

Then, $\Delta p = 20.80$ psf

So a transducer with a range from about 3 to 21 psf or 0.5 to 4 inches H$_2$O is needed. Electronic pressure transducers are available with a range from 0 to 5 inches H$_2$O, and such a transducer will be suitable. (Alternatively if a mechanical output is acceptable, a manometer would work well in this range at a fraction of the cost.)

Based on (9.13) and the propagation of uncertainty to a resultant, the uncertainty in flow rate can be written as,

$$u_Q = \pm \left[\left(\frac{\partial Q}{\partial C} u_C\right)^2 + \left(\frac{\partial Q}{\partial E} u_E\right)^2 + \left(\frac{\partial Q}{\partial A} u_A\right)^2 + \left(\frac{\partial Q}{\partial \rho} u_\rho\right)^2 + \left(\frac{\partial Q}{\partial \Delta p} u_{\Delta p}\right)^2 \right]^{1/2}$$

or dividing by (10.14), the uncertainty on a percent basis is

$$\frac{u_Q}{Q} = \pm \left[\left(\frac{u_C}{C}\right)^2 + \left(\frac{u_E}{E}\right)^2 + \left(\frac{u_A}{A}\right)^2 + \left(\frac{u_\rho}{2\rho}\right)^2 + \left(\frac{u_{\Delta p}}{2\Delta p}\right)^2 \right]^{1/2}$$

For the design-stage uncertainty analysis, we assume:

Transducer: $u_{\Delta p}/\Delta p = 0.0025$ [given]
Discharge coefficient: $u_C/C = 0.02$ [text ref. 2]
Diameters: $u_{d_1}/d_1 = u_{d_o}/d_o = 0.002$ [reasonable assumption]
Density: $u_\rho/\rho = 0.002$ [reasonable assumption]

Then,

Area: $u_A/A = 2u_{d_o}/d_o = 0.004$
Beta ratio: $u_\beta/\beta = [(u_{d_o}/d_o)^2 + (u_{d_1}/d_1)^2]^{0.5} = 0.003$
Approach factor: $u_E/E = [2\beta^3/(1-\beta^4)^2]u_\beta/\beta = 0.0009$

$$u_Q/Q = \pm [0.02^2 + 0.0009^2 + 0.004^2 + (0.002/2)^2 + (0.0025/2)^2]^{.5} = \pm 0.02$$

or about 2% of the flow rate.

COMMENT

Note how the known information about the instruments and the coefficients has been used to estimate this uncertainty. It is clear that the uncertainty in discharge coefficient dominates the uncertainty in flow rate. In situ calibration of the nozzle could reduce this uncertainty in C. But the uncertainty due to the pressure transducer is far too small to be an important factor in this measurement.

PROBLEM 10.19

KNOWN: Water at 60°F flowing through an orifice meter.
$10 \leq Q \leq 50$ gpm
$d_1 = 2.3$ inch
$u_{\Delta p}/\Delta p = \pm 0.005$
Select $C = f(\beta)$ only

FIND: d_o; design stage uncertainty in Q

ASSUMPTIONS: Steady, incompressible (Y = 1) flow of water.

PROPERTIES: Water: $\rho = 62.4$ lb$_m$/ft^3 $\nu = 1.2 \times 10^{-5}$ ft^2/s

SOLUTION

First, solve for d_o. From (10.14), $Q = CEAY(2\Delta p/\rho_1)^{.5}$ with $E = (1-\beta^4)^{-0.5}$. From Figure 10.6, K becomes essentially independent of Reynolds number at the higher values. So we will try to meet the constraint at the lowest expected flow rate so that it will be met at all higher flow rates.

At 10 gpm,

$$Re_{d1} = 4Q/\pi\nu d_1 = 4(10 \text{ gpm})(448 \text{ cfs/gpm})/\pi(1.2\times10^{-5} \text{ ft}^2/\text{s})(2.3/12 \text{ ft}) = 9900$$

Again, using Figure 10.6 at Re = 9900, select $\beta = 0.3$. The value for discharge coefficient falls within a range (C = K_o/E): $0.60 \leq C \leq 0.615$. Then with $d_o = 0.6$ inches, we can assume a constant value of C = 0.607 to within 2.5% over the flow range.

Based on (10.14) and the propagation of uncertainty to a resultant, and dividing by (10.14), the uncertainty in flow rate can be written as, (or see prob 10.18)

$$u_Q/Q = \pm [(u_C/C)^2 + (u_A/A)^2 + (u_E/E)^2 + (u_\rho/2\rho)^2 + (u_{\Delta p}/2\Delta p)^2]^{.5}$$

From Chapter 10 reference 2 or 12, we can expect an error in discharge coefficient of $e_1 = 0.006C$ when using the tabulated values. Together with the error introduced by assuming that C is constant over the full flow, we obtain,

$$u_C/C = ((e_1/C)^2 + (e_2/C)^2)^{0.5} = (0.006^2 + 0.025^2)^{0.5} = 0.026$$

Further, it is reasonable to expect a tolerance of 0.005 inch in all dimensional measurements. Hence,

$u_{do}/d_o = 0.0083$
$u_{d1}/d_1 = 0.0025$

Then,

$$u_A/A = 2\pi d_o u_{do}/4A = 2u_{do}/d_o = 0.0167$$

$$u_\beta/\beta = [(u_{do}/d_o)^2 + (u_{d1}/d_1)^2]^{0.5} = 0.017$$

$$u_E/E = [2\beta^3/(1-\beta^4)^2]u_\beta/\beta = 0.0009$$

It is reasonable to assume a ± 0.5% bias error in density, so

$$u_\rho/\rho = 0.005$$

Then with $u_{\Delta p}/\Delta p = 0.005$,

$$u_Q/Q = \pm[\,0.026^2 + 0.0167^2 + 0.0009^2 + (0.005/2)^2 + (0.005/2)^2\,]^{0.5}$$
$$= \pm 0.031$$

The uncertainty in flow rate is about 3.1%.

COMMENT

Certainly, the uncertainty in C dominates the uncertainty in Q. It is worth noting that as Re_{d1} approaches 50 000 and beyond (Q > 40 gpm), the value for C approaches 0.60. Using this value for all flow rates would retain the previous uncertainty value at low flow rates but would permit u_Q/Q to reduce to less than 1.8% at the higher flow rates.

PROBLEM 10.20

KNOWN: Air flow through an orifice
$10 \le \Phi \le 80\%$ where Φ is relative humidity
$\rho(\Phi = 45\%)$ used

FIND: u_Q due to variation in density

ASSUMPTIONS T = 70°F

PROPERTIES: Using a psychiometric chart for air at standard atmosphere,
$\rho(80\%) = 0.0746$ lb$_m$/ft^3
$\rho(10\%) = 0.0735$ lb$_m$/ft^3
$\rho(45\%) = 0.0741$ lb$_m$/ft^3

SOLUTION

Inspection of the properties reveals that the density can be estimated as

$$\rho = 0.0741 \pm 0.0005 \text{ lb}_m/\text{ft}^3$$

over the range of relative humidity values. This does not account for bias errors in the actual density values obtained from the chart.
From (10.14),

$$Q = CEAY(2\Delta p/\rho_1)^{.5}$$

To isolate the effects of humidity, we neglect all error sources except those in of density (as density is affected by humidity). Then,

$$u_Q/Q = f(\rho_1) = \pm [(u_{\rho 1}/2\rho_1)^2]^{.5} = 0.0034$$

or an uncertainty of about 0.34% of the flow rate results due to relative humidity effects alone.

PROBLEM 10.21

KNOWN: Air flow through an orifice meter with the conditions from Problem 10.20.
$Q = 17$ m^3/hr $= 17$ cmh
$T = 20°C$
$\beta = 0.4$
$d_1 = 6$ cm
$u_{\Delta p}/\Delta p = 0.005$
$u_{do} = u_{d1} = 0.1$ mm
$p_1 = 96.5$ kPa

FIND: Design stage uncertainty in flow rate

PROPERTIES $R = 28$ N-m/kg-K
$\nu = 1.6 \times 10^{-5}$ m^2/s
$k = 1.4$

SOLUTION

From (10.14), $Q = CEAY(2\Delta p/\rho_1)^{.5}$ with $d_o = \beta d_1 = 2.4$ cm. But $K_o = CE = f(\beta, Re)$. At 17 cmh, $Re_{d1} = 4Q/\pi\nu d_1 = 6260$. Then from Figure 10.6, $K_o(0.4, 6260) = 0.63$ (at least to within ± 0.005). To solve (10.14) for Δp, we assume $Y = 1$. This yields for $A = 0.00045$ m^2,

$$\Delta p = [0.0047 \text{ cms}/(.63)(0.00045 \text{m}^2)(1)]^2(1.15 \text{ kg/m}^3/2) = 157.85 \text{ N/m}^2$$

Using this value, $\Delta p/p_1 = 0.0016$. With $k = 1.4$, Figure 10.7 gives $Y = 1$. So $\Delta p = 157.85$ N/m^2. This is equivalent to about 1.6 cm H$_2$O.

Based on (10.14) and the propagation of uncertainty to a resultant, dividing by (10.14), the uncertainty in flow rate can be written as

$$u_Q/Q = \pm [(u_C/C)^2 + (u_A/A)^2 + (u_E/E)^2$$
$$+ (u_Y/Y)^2 + (u_\rho/2\rho)^2 + (u_{\Delta p}/2\Delta p)^2]^{.5}$$

The uncertainty in C is due both to the intrinsic error in using tabulated values, e_1, and due to our ability to read Figure 10.6, e_2. Setting $e_1 = 0.006C$ (see text) and $e_2 = 0.008C$ (based on an estimated resolution of ± 0.005 from the chart):

$$u_C/C = (0.006^2 + 0.008^2)^{0.5} = 0.01$$

Other values:

$$u_{d_o}/d_o = 0.01/2.4 = 0.004$$

$$u_{d_1}/d_1 = 0.01/6 = 0.0017$$

$$u_A/A = 2u_{d_o}/d_o = 0.008$$

$$u_\beta/\beta = [(u_{d_o}/d_o)^2 + (u_{d_1}/d_1)]^{0.5} = 0.0043$$

$$u_E/E = 2\beta^3(1-\beta^4)^{-2}u_\beta/\beta = 0.0009$$

$$u_Y/Y = 0.004\Delta p/p_1 = 1.3 \times 10^{-5} \quad \text{(see text and reference 2)}$$

$$u_{\Delta p}/\Delta p = 0.005$$

Lastly, the uncertainty in density will be due both to the error from the presumed humidity variations, e_1, and the error in the tabulated values, e_2. Taking $e_1/\rho = (0.0005/1.15) = 0.0004$ and $e_2/\rho = 0.005$ (i.e. about 0.5% of the tabulated value):

$$u_\rho/\rho = [0.0004^2 + 0.005^2]^{0.5} = 0.005$$

Then,

$$u_Q/Q = \pm\{.01^2 + .0009^2 + .008^2 + (1.3\times 10^{-5})^2 + (.005/2)^2 + (.005/2)^2\}^{.5}$$

$$= \pm 0.0133$$

Accounting for known uncertainties, the uncertainty is about 1.3% of the flow rate.

COMMENT

The assignment of uncertainty estimates requires time, some research into available information and good common sense. Keep in mind that the above estimate does not include the effects of control of operating conditions and the measurement procedure.

PROBLEM 10.22

KNOWN: Air flow at 70°F through an ASME Long Radius Nozzle
Q = 45 cfm
p_1 = 14.1 psia

FIND: p_2 and d_o required to assure choked flow at the throat.

ASSUMPTIONS: Air behaves as a perfect gas in this process

PROPERTIES: R = 53.3 ft-lb/lb$_m$-°R
k = 1.4

SOLUTION

From (10.16), the critical pressure ratio is given by

$$[p_2/p_1]_{cr} = (2/k+1)^{k/k-1} = (2/2.4)^{1.4/0.4} = 0.528$$

The critical pressure ratio sets the largest pressure possible and still choke the throat. Then, the critical downstream pressure is

$$[p_2]_{cr} = 0.528 p_1 = 7.45 \text{ psia}$$

So, $p_2 \leq 7.45$ psia to choke the flow.

Under choked flow conditions, the mass flow rate is at a maximum value. Equation 10.18 gives the mass flow rate at the critical pressure ratio which must then be the maximum mass flow through the nozzle. Rearranging (10.18), we solve for the throat area at critical conditions which must represent the largest throat area that can choke the flow,

$$A_{max} = \frac{\dot{m}/p_1}{(2RT)^{1/2} \left(\frac{k}{k+1}\right) \left(\frac{2}{k+1}\right)^{2/k-1}\right)^{1/2}}$$

$$= \frac{(45 \text{ ft}^3/\text{min})(1 \text{min}/60 \text{s})}{\left(2 \times 53.3 \frac{\text{ft-lb}}{\text{lb}_m °R} \times 530°R \times 32.2 \frac{\text{lb}_m\text{-ft}}{\text{lb-s}^2}\right)^{1/2} \left(\frac{1.4}{2.4} \times \left(\frac{2}{2.4}\right)^{2/.4}\right)^{1/2}}$$

$$= 0.0011 \text{ ft}^2$$

But A = $\pi d_o^2/4$. So $d_o \leq 0.459$ inches would provide a suitable design.

PROBLEM 10.23

KNOWN: Water at 20°C flows through orifice plate with flange taps.
$d_1 = 10$ cm
$H = 100$ mm Hg at $Q = 2$ m^3/min $= 0.033$ cms

FIND: d_o required

SOLUTION

We want to have $H = 100$ mm Hg or $\Delta p = (13.57)(9800 \text{N/m}^2)(0.1 \text{ m}) = 13{,}300$ N/m^2. Equation 10.14 can be rewritten as:

$$d_o = [4Q/\{CE\pi(2\Delta p/\rho_1)^{.5}\}]^{.5}$$

for the conditions known, this reduces to $d_o = 0.09/(K_o)^{0.5}$ [m]

To solve, we that $K_o = CE = f(\beta, Re_{d1})$. For the target flowrate,

$$Re_{d1} = 4Q/\pi d_1 \nu = 4(0.033 \text{ cms})/\pi(.1\text{m})(1\times 10^{-6}\text{m}^2/\text{s}) = 425{,}000$$

Inspection of Figure 10.6 shows that the orifice diameter requires a β of nearly 1 to satisfy the flow meter equation. So an orifice meter can not be used to meet the given constraints!

PROBLEM 10.24

KNOWN: Nozzle used to meter 20°C water.
$d_1 = 20$ cm
$5000 \text{ cm}^3/\text{s} \leq Q \leq 50,000 \text{ cm}^3/\text{s}$
$\beta = 0.5$

FIND: Pressure transducer range required.

SOLUTION

Rewritting (10.14) with $Y = 1$ (incompressible):

$$\Delta p = (\rho/2)(Q/K_o A)^2$$

where $K_o = f(\beta = 0.5, Re_{d_1})$.

Low Q: $Re_{d_1} = 4(.005 \text{ m}^3/\text{s})/\pi(.2\text{m})(1 \times 10^{-6}/2/\text{s}) \approx 32,000$

High Q: $Re_{d_1} = 4(.05 \text{ m}^3/\text{s})/\pi(.2\text{m})(1 \times 10^{-6}/2/\text{s}) \approx 320,000$

So from Figure 10.11: $(K_o)_{low} \approx 1.000$; $(K_o)_{hi} = 1.005$.

Solving:

$(\Delta p)_{low} = 203 \text{ N/m}^2$

$(\Delta p)_{hi} = 20,100 \text{ N/m}^2$

So the selected transducer must have a range extending from about 200 to 20,500 N/m², one rated from 0 to 21 cm H₂O will work well.

PROBLEM 10.26

KNOWN: Vortex meter metering fluid flow

$St = 0.20$ (shedder Strouhal number)
$f = 77$ Hz (shed frequency)
$d = 1.27$ cm (shedder characteristic length, see Table 10.1)

FIND: Average velocity, $\overline{U}$

SOLUTION

Direct use of (10.27) gives,

$$St = \omega d / \overline{U} = 2\pi f d / \overline{U}$$

$$\overline{U} = 2\pi(77 \text{ Hz})(0.0127)/0.2 = 30.72 \text{ m/s}$$

PROBLEM 10.27

KNOWN: Air flow at 30°C metered by a thermal mass flow meter
$d_1 = 2$ cm
Power $= P = 25$ W
$\Delta T = 1°C = 1K$
$c_p = 1.006$ kJ/kg-K

FIND: mass flow rate

ASSUMPTIONS: c_p is constant through meter and known.
ΔT reported (measured) must be the average mixed temperature across each pipe cross section.
Power supplied to meter is 100% dissipated in fluid (i.e. no losses).

SOLUTION

From (10.33), $\dot{E} = \dot{m} c_p \Delta T$

But the energy supplied to the meter is dissipated into the flow, so that $\dot{E} = P$. Then,

$$\dot{m} = P/c_p \Delta T = (25 \text{ W})(1006 \text{ J/kg-K})(1K) = 0.025 \text{ kg/s}$$

COMMENT

Note the many assumptions that go into using this meter. Actually, for flows of gases which can be well modeled as perfect gases, these types of meters can provide excellent accuracy. For other types of fluids, they have limited applicability.

PROBLEM 10.28

KNOWN: Sonic nozzle used to regulate volume flow rate of air.

FIND: Uncertainty due to normal changes in ambient temperature and pressure.

SOLUTION

From (10.18),

$$\dot{m} = \rho_1 A (2RT_1)^{.5} [(k/k+1)(2/k+1)^{2/k+1}]^{.5}$$

For air, $(p_2/p_1)_{cr} = 0.528$. Provided that the actual p_2/p_1 ratio remains below this value the throat will remain choked. From (10.3) and (10.6), we can rewrite this for volume flow rate as,

$$Q = A(2RT_1)^{.5} [(k/k+1)(2/k+1)^{2/k+1}]^{.5}$$

This shows that the volume flow rate will be affected by environmental changes in temperature even though constant mass flow rate is maintained in the nozzle throat. If we look only at the effects of temperature variation over the ±25 °R range specified, then $Q = f(T)$ only. This can expressed as

$$u_Q/Q = [(u_T/2T)^2]^{0.5}$$
$$= [(25°R/2(510°R))^2]^{0.5}$$
$$= 0.0245$$

or an added uncertainty in volume flow rate of 2.45% due to temperature changes alone.

PROBLEM 10.29

KNOWN: From Example 10.4: Q = 0.053 cms (air)
d_1 = 6 cm
β = 0.4
d_o = 2.4 cm
H = 250 cm H_2O
p_1 = 93.7 kPa
T_1 = 293 K
Y = 0.92

All dimensions known to ±0.025 mm
Upstream pressure is constant
Pressure drop has bias of ±0.25 cm H_2O
S_H = 0.5 cm H_2O with N = 20

FIND: u_Q

PROPERTIES: water: γ = 9800 N/m³; ρ = 998 kg/m³
air: ν = 1.6 x 10⁻⁵ m²/s; ρ = 1.16 kg/m³

SOLUTION

From (10.14), the flow rate is found by

$$Q = f(C,E,A,Y,\Delta p,\rho)$$

with $Q = CEAY(2\Delta p/\rho_1)^{.5}$
and
$$\Delta p = (9800 \text{ N/m}^3)(2.5 \text{ m } H_2O) = 24.5 \text{ kPa}$$

The bias error in Q can be expressed on a percent basis as,

$$B_Q/Q = \pm[(B_C/C)^2+(B_E/E)^2+(B_A/A)^2+(B_Y/Y)^2 + (B_\rho/2\rho)^2+(B_{\Delta p}/\Delta p)^2]^{.5}$$

where

B_{do}/d_o = 0.0025/2.4 = 0.001

B_{d1}/d_1 = 0.0025/6 = 0.0004

$B_\beta = [(B_{do}/d_o)^2 + (B_{d1}/d_1)^2]^{0.5}$ = 0.001

$$B_A/A = 2B_{do}/d_o = 0.002$$

$$B_E/E = 2\beta^3(1-\beta^4)^{-2}B_\beta/\beta = 0.0002$$

$$B_Y/Y = 0.004\Delta p/p_1 = 1.3 \times 10^{-5} \quad \text{(see text and reference 2)}$$

$$B_{\Delta p}/\Delta p = 0.25/250 = 0.001$$

$$B_\rho/\rho = 0.005 \quad \text{(assumed value)}$$

According to the text and reference 2, the bias error in using the tabulated values for C amounts to 0.006C. However, we will also add an additional elemental bias error in reading Figure 10.6 of, say, 0.008C. Then,

$$B_C/C = (0.006^2 + .008^2)^{0.5} = 0.0094$$

Inserting yields,

$$B_Q/Q = [0.0094^2 + 0.0002^2 + 0.002^2 + 0.0000013^2 + (0.005/2)^2 + (0.001/2)^2]^{.5}$$
$$= 0.01$$

Or $B_Q = 0.00053$ m^3/s. Note how the bias in C dominates.

Likewise the precision error in Q can be expressed as

$$P_Q/Q = \pm[(P_C/C)^2 + (P_E/E)^2 + (P_A/A)^2 + (P_Y/Y)^2 + (P_\rho/2\rho)^2 + (P_{\Delta p}/\Delta p)^2]^{.5}$$

But the only precision error estimate comes from the pressure reading, so that this expression can be reduced to

$$P_Q/Q = \pm P_{\Delta p}/2\Delta p$$

The precision in pressure reading is estimated from

$$S_H = 0.5 \text{ cm H}_2\text{O}$$

The precision in the mean value of the pressure reading is estimated at

$$P_H = S_{\bar{H}} = S_H/(20)^{0.5} = 0.11 \text{ cm H}_2\text{O}$$

or in correct units,

$$P_{\Delta p} = 9.8 \text{ N.m}^2$$

with a degrees of freedom of 19. Then,

$$P_Q/Q = [(P_{\Delta p}/2\Delta p)^2]^{0.5}$$
$$= [(9.8/(2)(24500))^2]^{0.5} = 0.0002$$

Or $P_Q = 1 \times 10^{-5}$ m^3/s.

Combining bias and precision estimates for flow rate,

$$u_Q = \pm[0.00053^2 + (2.093 \times 1 \times 10^{-5})^2]^{0.5} = \pm 0.00053 \text{ cms } (95\%)$$

COMMENT

The uncertainty amounts to about 1% of the flow rate and is primarily due to bias errors in the variables, notably the discharge coefficient. The user may feel more comfortable using larger values for the bias errors obtained from variable read from charts and properties obtained from tables. The point is that a careful consideration of each term relevant to the measurement has been made.

PROBLEM 11.1

KNOWN: A steel rod (circular cross-section) having:

$L = 10$ in. $\quad E_M = 30 \times 10^6 \; lb/in^2$

$D = 0.25$ in $\quad m = 40 \; lb_m$

FIND: The change in length of the rod, δL

ASSUMPTIONS: Rod is elastically deformed, such that

$$\sigma = \varepsilon E_m$$

SOLUTION:

The force resulting from 40 lb_m in standard gravity is

$$F_n = \frac{ma}{g_c} = \frac{(40 \; lb_m)(32.174 \; ft/sec^2)}{32.174 \; \frac{ft-lb_m}{lb-sec^2}} = 40 \; lb$$

The resulting uniaxial stress is

$$\sigma_a = \frac{F_n}{A_c}$$

where

$$A_c = \frac{\pi}{4}(0.25)^2 = 0.049 \; in.^2$$

$$\sigma_a = \frac{40 \; lb}{0.049 \; in^2} = 814.9 \; lb/in^2$$

and

$$\varepsilon_a = \frac{\sigma_a}{E_M} = \frac{814.9 \; lb/in^2}{30 \times 10^6 \; lb/in^2} = 2.716 \times 10^{-5}$$

The change in length is then

$$\delta L = L \varepsilon_a = (2.716 \times 10^{-5})(10 \; in.) = 0.00027 \; in.$$

PROBLEM 11.2

KNOWN: A steel rod (circular cross-section) having:

$$L = 0.3 \, m$$
$$D = 5 \, mm$$
$$E_m = 20 \times 10^{10} \, Pa$$
$$m = 50 \, kg$$

FIND: The change in length of the rod, ∂L.

ASSUMPTIONS: Rod is elastically deformed, such that

$$\sigma = \varepsilon \cdot E_m$$

SOLUTION:

The force resulting from 50 kg in standard gravity is

$$F_N = \frac{ma}{g_c} = \frac{(50 \, kg)(9.8 \, m/s^2)}{1.0 \, \frac{kg-m}{s^2-N}}$$

$$F_N = 490 \, N$$

The resulting uniaxial stress is

$$\sigma_a = \frac{F_N}{A_c}$$

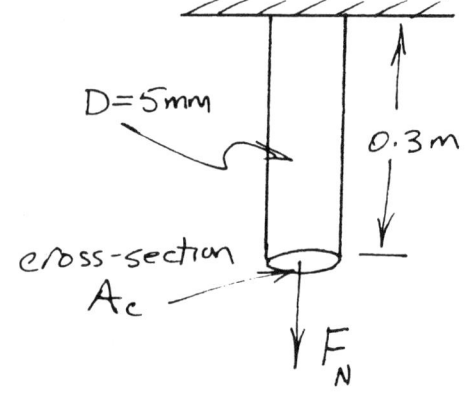

where

$$A_c = \frac{\pi}{4}(5 \times 10^{-3})^2 = 1.96 \times 10^{-5} \, m^2$$

$$\sigma_a = \frac{490 \, N}{1.96 \times 10^{-5} \, m^2} = 25 \times 10^6 \, Pa$$

and

$$\varepsilon_a = \frac{\sigma_a}{E_m} = \frac{25 \times 10^6 \, Pa}{20 \times 10^{10} \, Pa} = 125 \times 10^{-6}$$

The change in length is then

$$\partial L = L \cdot \varepsilon_a = (0.3)(125 \times 10^{-6}) = 37.5 \times 10^{-6} \, m$$

PROBLEM 11.3

KNOWN: An electrical coil with

$N = 20,000$
$D = 0.051$ in.
$r = 2.0$ in.

FIND: The resistance, R.

SOLUTION:

We know

$$R = \frac{\rho_e L}{A_c}$$

and

$$A_c = \frac{\pi}{4}(D)^2 = \frac{\pi}{4}\left(\frac{0.051}{12}\right)^2 = 1.42 \times 10^{-5} \text{ ft}^2$$

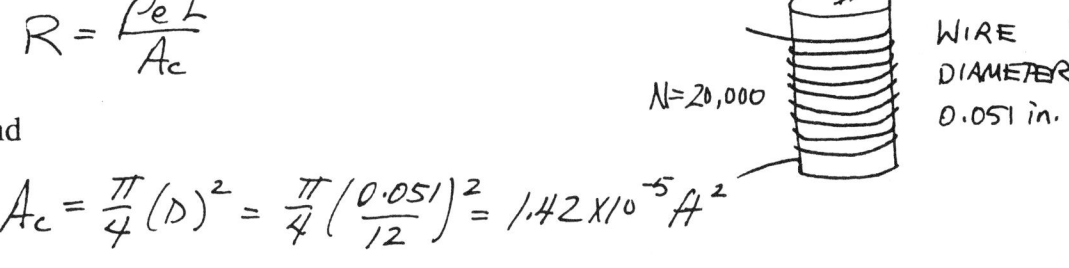

The length is then found as

$$L = 2\pi r (N) = 2\pi \left(\frac{2}{12}\right)(20,000) = 20,944 \text{ ft}$$

which yields a resistance of

$$R = \frac{(1.673 \times 10^{-6} \ \Omega\text{-cm})(0.0328 \text{ ft/cm})(20,944 \text{ ft})}{1.42 \times 10^{-5} \text{ ft}^2}$$

$$R = 81 \ \Omega$$

PROBLEM 11.4

KNOWN: Aluminum having a volume of 3.14159 x 10^{-5} m^3, and a resistivity, ρ_e = 2.66 x 10^{-8} Ω-m.

FIND: Resistance, R, of 2 mm and 1 mm diameter wires having the same total volume.

SOLUTION:

Volume for a cylindrical wire is

$$V = \frac{\pi}{4}(D^2)L$$

yielding

L_{2mm} = 10 m and L_{1mm} = 40 m

The resistance values are then calculated as

$$R_{2mm} = \frac{\rho_e L}{A_c} \qquad A_c = \frac{\pi}{4}D^2$$

$$R_{2mm} = \frac{(2.66 \times 10^{-8}\,\Omega\text{-m})(10\,m)}{\frac{\pi}{4}(2 \times 10^{-3})^2} = 0.085\,\Omega$$

$$R_{1mm} = \frac{(2.66 \times 10^{-8}\,\Omega\text{-m})(40\,m)}{\frac{\pi}{4}(1 \times 10^{-3})^2} = 1.355\,\Omega$$

PROBLEM 11.5

KNOWN: A nickel conductor with

$$\rho_e = 6.8 \times 10^{-8} \ \Omega\text{-m}$$

$A_c = 5 \times 2$ mm (rectangular)

$L = 5$ m

FIND: a) R - the total resistance

b) The diameter of a 5 m long copper wire having a circular cross-section to yield the same resistance.

SOLUTION:

The resistance is found from

$$R = \frac{\rho_e L}{A_c} \qquad A_c = 10 \ mm^2$$

$$L = 5m$$

which in this case yields

$$R = \frac{(6.8 \times 10^{-8} \ \Omega\text{-m})(5m)}{10 \times 10^{-6} \ m^2} = 0.034 \ \Omega$$

For the copper, $\rho_e = 1.7 \times 10^{-8} \ \Omega\text{-m}$

$$0.034 = \frac{(1.7 \times 10^{-8} \ \Omega\text{-m})(5m)}{A_c}$$

$A_c = 2.5 \times 10^{-6} \ m^2 \quad D = 1.8$ mm

PROBLEM 11.6

KNOWN: A Wheatstone bridge with all resistances initially equal to 100 Ω

The maximum power through R_1 is 0.25 W.

FIND: Maximum applied voltage
Bridge sensitivity

ASSUMPTIONS: Infinite meter resistance

SOLUTION: From the circuit shown below
$i_1 R_1 + i_2 R_2 = E_i$
$i_1 = i_2 = i$
$i(R_1 + R_2) = E_i$

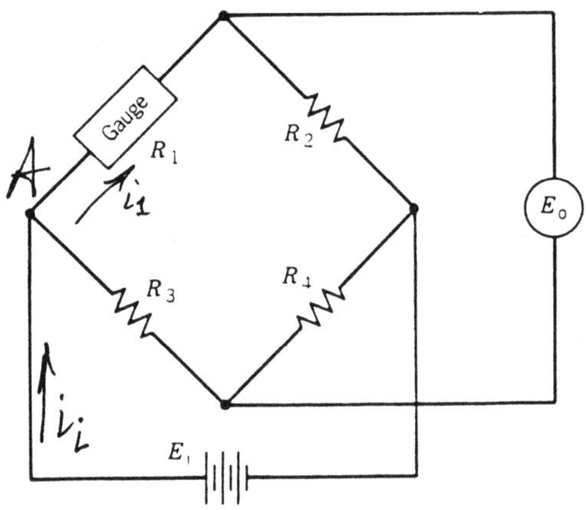

But we know power, P, is given by $P = i^2 R$ and
$i = \sqrt{\dfrac{0.25 \text{ W}}{100 \, \Omega}} = 0.05 \text{A}$

At node A

$i_i = i_1 + i_3 = 2i \quad i_i = 0.1 \text{ A}$

$E_i = i_i R_B$ where R_B is the equivalent bridge resistance

so $E_i = (0.1 \text{ A})(100 \, \Omega) = 10 \text{ V}$

The bridge sensitivity is defined as

$$K_B = \frac{\delta E_o}{\delta R_1}$$

and for a bridge with all resistances initially equal and assuming $\delta R \gg R$

$$\frac{\delta E_o}{\delta R} \approx \frac{E_i R}{4R} = \frac{(10 \text{ V})(100 \, \Omega)}{400 \, \Omega} \approx 2.5 \text{ V}/\Omega$$

PROBLEM 11.7

KNOWN: A strain gauge with $R_1 = 120$, $GF = 2$ in an equal arm Wheatstone bridge $R_2 = R_3 = R_4 = 120\ \Omega$. Maximum gauge current is 0.05 A

FIND: Maximum input bridge voltage

SOLUTION: From a basic circuit analysis, assuming infinite meter resistance

$$i_1 = \frac{E_i}{R_1 + R_2}$$

and

$$E_i = i_1(R_1 + R_2)$$
$$E_i = (0.05\ A)(240\ \Omega) = 12\ V$$

PROBLEM 11.8

KNOWN: A strain gauge has a nominal resistance of 350 Ω, and GF = 1.8, and senses axial strain. The gauge is mounted on a 1 cm^2 aluminum rod (Em = 70 GPa)

$$E_o = 1 \text{ mV}, \quad E_i = 5 \text{ V}$$

FIND: Applied load, assuming uniaxial tension

SOLUTION:

For an equal arm bridge, from (11.14)

$$\frac{\delta E_o}{E_i} = \frac{(\delta R/R)}{4 + 2(\delta R/R)}$$

$$\frac{0.001}{5} = \frac{\delta R/R}{4 + 2 \delta R/R}$$

and

$$\delta R/R = 0.0008 \qquad \delta R = 0.28 \, \Omega$$

Since

$$\delta R/R = \varepsilon \cdot GF \qquad \varepsilon = 0.00044$$

and with

$$\varepsilon = \sigma/E_m \qquad \sigma = \varepsilon E_m$$
$$= (0.00044)(70 \text{ GPa})$$

$$\sigma = 0.0308 \text{ GPa}$$

To find the applied force, F_N

$$\sigma = F_N / A_c \qquad A_c = \frac{\pi}{4}(1 \times 10^{-2} m)^2 = 7.854 \times 10^{-5} m^2$$

$F_N = (7.854 \times 10^{-5} m^2)(0.0308 \times 10^9 Pa)$

and since

$$1 Pa = 1 N/m^2$$

$$F_N = 2419 N$$

COMMENT: This force would result from a mass, in standard gravitational acceleration of 246.6 kg.

PROBLEM 11.9

KNOWN: Strain gauge installation shown in Figure 11.13.

FIND: Show that this installation is not sensitive to bending stresses.

SOLUTION:
The stresses created by a bending and axial load may be represented as

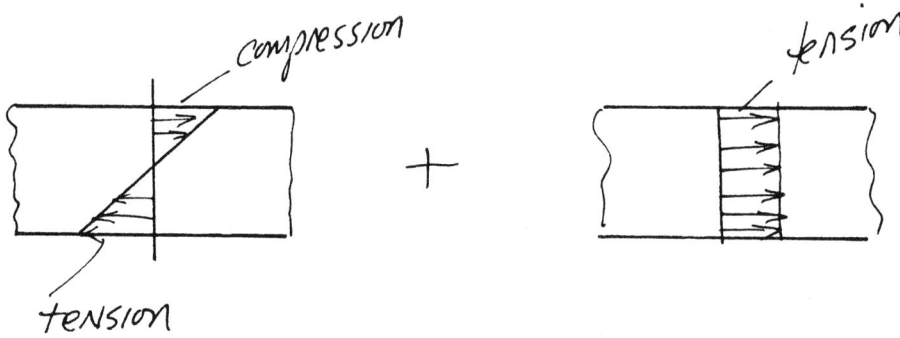

In general, for four active elements in a bridge,

$$\frac{\delta E_o}{E_i} = \frac{GF}{4}(\varepsilon_1 - \varepsilon_2 + \varepsilon_4 - \varepsilon_3)$$

or for this case

$$\frac{\delta E_o}{E_i} = \frac{GF}{4}(\varepsilon_1 + \varepsilon_4)$$

Strain may be expressed as a linear combination of the imposed loads

$$\frac{\delta E_o}{E_i} = \frac{GF}{4}\left[(\varepsilon_M + \varepsilon_{F_N})_1 + (\varepsilon_M + \varepsilon_{F_N})_4\right]$$

But since $\varepsilon_{M_1} = -\varepsilon_{M_4}$ this installation is not sensitive to bending.

PROBLEM 11.10

KNOWN: A steel member (ν_p = 0.3) subject to simple axial tension. Strain gauges are mounted on top center, and bottom center.

FIND: Bridge constant, for gauge locations 1 and 4.

SOLUTION:

The configuration is

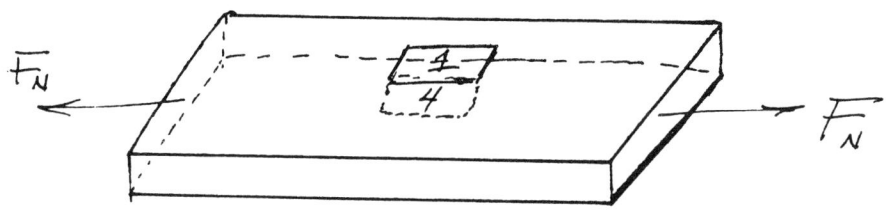

Since for any 4 gauges

$$\frac{\partial E_o}{E_i} = \frac{GF}{4}(\varepsilon_1 - \varepsilon_2 + \varepsilon_4 - \varepsilon_3)$$

and for a single gauge, sensing maximum strain

$$\frac{\partial E_o}{E_i} = \frac{GF}{4}(\varepsilon_{max})$$

In the present case both gauges sense the maximum strain, and the outputs are additive

$$K = 2$$

∂E_o = 10 mV and E_i = 10 V R = 120 Ω and , GF = 2

For the gauge sensing maximum strain

$$\frac{\delta E_o^s}{E_i} = \left[\frac{\varepsilon_{max} GF}{4 + 2\varepsilon_{max} GF}\right]$$

The actual output is then

$$\delta E_o = K \delta E_o^s$$

Solving for ε_{max} yields

$$\varepsilon_{max} = 0.001$$

Therefore

$$\varepsilon_{axial} = \varepsilon_{max} = 0.001$$

$$\varepsilon_t = 0.0003$$

PROBLEM 11.11

KNOWN: An axial and a transverse strain gauge are mounted to the top surface of a steel beam, and connected in arms 1 and 2 of a Wheatstone bridge.

$\partial E_o = 250 \mu V$ $\nu_p = 0.3$ $\sigma = 2222.2 \, psi$

$E_i = 10 V$ $E_m = 29.4 \times 10^6 \, psi$

FIND: a) K_B

b) Average gauge factor

SOLUTION:

In general

$$\frac{\partial E_o}{E_i} = \frac{GF}{4}(\varepsilon_1 - \varepsilon_2 + \varepsilon_4 - \varepsilon_3)$$

which in this case, for one axial and one transverse gauge yields

$$\frac{\partial E_o}{E_i} = \frac{GF}{4}(\varepsilon_{max} - 0.3 \varepsilon_{max})$$

$$= \frac{GF}{4}(0.7 \varepsilon_{max})$$

for a single gauge sensing the maximum strain

$$\frac{\partial E_o}{E_i} = \frac{GF}{4} \varepsilon_{max}$$

AND $K_B = 0.7$

To find the average value of GF

$$\frac{\delta E_o}{E_i} = \frac{GF}{4} \varepsilon_{max} K_B$$

$$\frac{250 \times 10^{-6}}{10} = \frac{GF}{4} \varepsilon_{max} K_B$$

with $\sigma = 2222.2$ psi and $\varepsilon_{max} = \sigma/E_m$

$$\varepsilon_{max} = 7.559 \times 10^{-5}$$

and

$$GF = \frac{4(250 \times 10^{-6})}{10 \,(0.7)(7.559 \times 10^{-5})} = 1.89$$

Comment: The chosen arrangement of strain gauges yields a bridge constant less than one, which without other considerations, is not a good choice.

PROBLEM 11.12

KNOWN: A strain gauge, mounted on a steel cantilever, has the following characteristics:

$R = 120\ \Omega$

$\delta R = 0.1\ \Omega$

$GF = 2.05 \pm 1\%\ (95\%)$

$u_R = \pm 1\%$

FIND: Estimate the strain, ε_a, and the uncertainty in the measured strain, u_ε.

ASSUMPTIONS: The bridge operates in a null mode and reasonable values for input voltage and galvanometer sensitivity must be assigned.

SOLUTION:

For an equal arm bridge,

$$\frac{\delta R_1}{R_1} = \varepsilon_a \cdot GF$$

which yields

$$\frac{0.1}{120} = \varepsilon_a (2.05) \qquad \varepsilon_a = 0.000407$$

The uncertainty analysis can be approached in several ways. Since the bridge is operated in a null mode, a galvanometer and a calibrated resistor are employed. The following relationships are used for the bridge

$$\frac{\delta E_o}{E_i} = \frac{(\delta R/R)}{4 + 2(\delta R/R)}$$

$$R_1 = R_2 \left(\frac{R_3}{R_4} \right) \quad \text{(balanced bridge)}$$

Let $\quad \gamma = \left(\delta R_1 / R_1 \right)$

A typical galvanometer sensitivity may be $\pm 1\ \mu V$, and $E_i = 10\ V$, then

$$\varepsilon = \gamma / GF$$

$$u_\varepsilon = \left[\left(\frac{\partial \varepsilon}{\partial \gamma} u_\gamma\right)^2 + \left(\frac{\partial \varepsilon}{\partial GF} u_{GF}\right)^2\right]^{1/2}$$

This equation for the uncertainty in strain contains two uncertainties, yet to be estimated. The partial derivatives which represent the sensitivity indices are evaluated at the nominal values as

$$\frac{\partial \varepsilon}{\partial \gamma} = \frac{1}{GF} = \frac{1}{2.05}$$

$$\frac{\partial \varepsilon}{\partial GF} = -\frac{\gamma}{GF^2} = -\frac{0.1}{(2.05)^2(120)} = 2 \times 10^{-4}$$

We must examine the uncertainty in γ, which has contributions from the galvanometer, and from the bridge and calibrated resistors.
The analysis proceeds as

$$\gamma = \left[\frac{4(\delta E_o / E_i)}{1 + 2(\delta E_o / E_i)}\right]$$

$$u_\gamma^b = \frac{\partial \gamma}{\partial(\delta E_o / E_i)} u_{\delta E_o / E_i} \qquad \text{AT } \delta E_o = 0$$

$$\frac{\partial \gamma}{\partial(\delta E_o / E_i)} = 4$$

ASSUME THE ONLY CONTRIBUTIONS TO $\delta E_o / E_i$ IS THE GALVANOMETER, $\pm 1 \mu V \Rightarrow u_{\delta E_o / E_i} = \pm 1 \times 10^{-7} V$

ADDITIONAL CONTRIBUTIONS TO UNCERTAINTY IN γ RESULT FROM

γ — BRIDGE RESISTANCES $R_3, R_4 \pm 1\%$
  — CALIBRATED RESISTOR $R_2 \pm 1\%$

AND WITH $R_1 = R_2 \left(R_3/R_4\right)$

$$\frac{\partial R_1}{\partial R_2} = \frac{R_3}{R_4} = 1$$

$$\frac{\partial R_1}{\partial R_3} = \frac{R_2}{R_4} = 1$$

$$\frac{\partial R_1}{\partial R_4} = -\frac{R_2 R_3}{R_4^2} = 1$$

$$U_{R_1} = \sqrt{3(0.01)^2} = \pm 0.0173 \,\Omega$$

$$U_\gamma^a = \frac{0.0173}{120} = \pm 0.000144 \,\Omega$$

COMBINING U_γ^a and U_γ^b

$$U_\gamma = \sqrt{[4(1\times10^{-7})]^2 + [0.000144]^2} = \pm 0.000144$$

THEN WITH

$$U_\varepsilon = \left[\left(\frac{1}{GF} U_\gamma \right)^2 + \left(-\frac{\gamma}{GF^2} U_{GF} \right)^2 \right]^{1/2}$$

$$= \left[\left(\frac{1}{2.05} \, 0.000144 \right)^2 + \left(-\frac{0.1/120}{2.05^2} \, 0.02 \right)^2 \right]^{1/2}$$

With this result, the uncertainty in strain is found as

$$\underline{\underline{U_\varepsilon = \pm 7 \times 10^{-5}}}$$

PROBLEM 11.13

KNOWN:
R = 120 Ω   Gauges mounted on opposite arms of bridge
E_i = 4 V E_o = 120 μV
GF = 2 E_m = 29 x 10^6 psi

FIND: Resistance change for each gauge

SOLUTION:

For this bridge

$$\frac{\partial E_o}{E_i} = \frac{GF}{4}\left(\varepsilon_1 - \cancel{\varepsilon_2}^{0} + \varepsilon_4 - \cancel{\varepsilon_3}^{0}\right)$$

which implies a bridge constant of 2.

Thus

$$\frac{\partial E_o}{E_i} = \frac{GF}{4}(2\varepsilon_a) = \frac{K_B \, GF}{4}(\varepsilon_{max})$$

$$\varepsilon_{max} = \left(\frac{120 \times 10^{-6}}{4}\right)\left(\frac{4}{2(2)}\right) = 3.0 \times 10^{-5}$$

Since

$$\frac{\partial R}{R} = \varepsilon_{max} \, GF \qquad \partial R = (120\,\Omega)(3\times 10^{-5})(2)$$

The change in resistance is

$$\partial R = 0.0072 \, \Omega$$

PROBLEM 11.14

KNOWN: A strain gauge Wheatstone bridge circuit with
Galvanometer resolution: 1 microamp
Galvanometer R: 100 Ω   ± 0.5% (95%)
$R_2 = R_3 = 120$ Ω   ± 1% (95%)
R_p known to within 1% (95%)
GF = 2
$E_i = 4$ V
Galvanometer current = 2×10^{-6} A

FIND: The indicated strain and its uncertainty

SOLUTION:
The initial value of R_1 (the gauge resistance) is found from the condition for a balanced bridge as 120.07 Ω.
Equation 6.28 can be used to solve for a new value of R_1 as

$$I_g = \frac{E_i(R_3 R_2 - R_1 R_4)}{R_3(R_1+R_2)(R_g+R_2+R_4) + R_1 R_2 R_4 - R_3 R_2^2 + R_g R_4(R_1+R_2)}$$

which yields $R_1 = 119.8773$, or $\delta R = -0.193$ Ω
The strain is found from

$$\frac{\delta R}{R} = \varepsilon \cdot GF \qquad \varepsilon = 0.0008$$

The uncertainty in strain may be estimated at the design stage as

$$u_\varepsilon = \left[\left(\frac{\partial \varepsilon}{\partial GF} u_{GF}\right)^2 + \left(\frac{\partial \varepsilon}{\partial (\delta R/R)} u_{\delta R/R}\right)^2 \right]^{1/2}$$

Thus we need an uncertainty estimate for $\delta R/R$
Taking the uncertainties in the bridge resistances to be 1.2 Ω
(with bridge resistances approximately equal, the sensitivity indices are 1, see solution for problem 11.12)

$$U_{R_1} = \sqrt{3(1.2)^2} = \pm 2.1 \, \Omega$$

The contribution to the uncertainty in $\partial R/R$ from the meter resistance can be estimated from equation 6.30

$$I_g = E_i \frac{\partial R/R}{4(R+R_g)}$$

Thus the meter contributions can be combined as

$$U_{\partial R/R} = \left[\left(\frac{\partial (\partial R/R)}{\partial I_g} U_{I_g} \right)^2 + \left(\frac{\partial (\partial R/R)}{\partial R_g} U_{R_g} \right)^2 + \left(\frac{U_{R_1}}{R_1} \right)^2 \right]^{1/2}$$

With sensitivities

$$\frac{\partial (\partial R/R)}{\partial I_g} = \frac{4}{E_i}(R+R_g) = 220$$

$$\frac{\partial (\partial R/R)}{\partial R_g} = 2 \times 10^{-6}$$

$$U_{\partial R/R} = \frac{U_{R_1}}{R_1} = \frac{\pm 2.1}{120} = \pm 0.017 \, \Omega$$

Then, assuming no uncertainty in the gauge factor,

$$U_\varepsilon = \frac{1}{GF} U_{\partial R/R} = \pm 0.0085$$

$$\varepsilon = 0.0008 \pm 0.0085 \; (95\%)$$

COMMENT: This uncertainty analysis clearly shows that a measurement of strain using this bridge would not be acceptable, and points to the need for a more precise knowledge of the bridge resistance values.

PROBLEM 11.15

KNOWN: Bridge arrangements of Figure 11.23

FIND: Bridge constants

SOLUTION:

Using equation 11.25

$$\frac{\delta E_o}{E_i} = \frac{GF}{4}(\varepsilon_1 - \varepsilon_2 + \varepsilon_4 - \varepsilon_3)$$

For a single gauge sensing the maximum strain

$$\frac{\delta E_o}{E_i} = \frac{GF}{4}\varepsilon_{max}$$

a)

$$K_B = \frac{\frac{GF}{4}\varepsilon_1}{\frac{GF}{4}\varepsilon_{max}} = 1$$

b)

$$K_B = \frac{\frac{GF}{4}(\varepsilon_1 - \varepsilon_3)}{\frac{GF}{4}\varepsilon_{max}} = \frac{\frac{GF}{4}[\varepsilon_1 - (-\nu_p\varepsilon_1)]}{\frac{GF}{4}\varepsilon_1}$$

$$K_B = 1 + \nu_p$$

c)

$$K_B = \frac{\frac{GF}{4}(\varepsilon_1 - \varepsilon_3)}{\frac{GF}{4}\varepsilon_1} = \frac{\frac{GF}{4}[\varepsilon_1 - (-\varepsilon_1)]}{\frac{GF}{4}\varepsilon_1} = 2$$

d)
$$K_B = \frac{\frac{GF}{4}(\varepsilon_1 - \varepsilon_2 + \varepsilon_4 - \varepsilon_3)}{\frac{GF}{4}\varepsilon_1}$$

$\varepsilon_2 = -\nu_p \varepsilon_1 \quad \varepsilon_3 = -\nu_p \varepsilon_4$

$K_B = 2(1 + \nu_p)$

e)
$$K_B = \frac{\varepsilon_1 - \varepsilon_2 + \varepsilon_4 - \varepsilon_3}{\varepsilon_1}$$

$\varepsilon_2 = -\varepsilon_1 \qquad K_B = 4$

$\varepsilon_3 = -\varepsilon_4$

PROBLEM 11.16

KNOWN: D = 1 m, Tranverse sensitivity = 0.03 = 3%.

FIND: Error due to tranverse sensitivity.

SOLUTION:
Since

$$\sigma_t = \frac{PD}{2t} \quad \text{and} \quad \sigma_l = \frac{PD}{4t} \quad \text{then} \quad \frac{\varepsilon_t}{\varepsilon_a} = \frac{\sigma_l}{\sigma_t} = \frac{\frac{PD}{4t}}{\frac{PD}{2t}} = \frac{1}{2}$$

From Figure 11.6, the error can be determined as 2.8% of σ_t.

PROBLEM 11.17

KNOWN: Wheatstone bridge circuit, with all fixed resistances equal to 100 Ω.
R_1 is a strain gauge, with a resistance of 100 Ω at zero strain. The strain gauge senses ε_1.
t = 2 cm   D = 2 m   GF = 2   Maximum power dissipation in gauge = 0.25 W

FIND: Maximum allowable static sensitivity, in V/kPa
Under what conditions is K constant.

SOLUTION:

From Problem 11.6, the maximum value of E_i to limit power dissipation to 0.25 W is 10 V, and the static sensitivity is 2.5 V/Ω, based on the resistance change of the strain gauge. In the present problem, we can write

$$\sigma_1 = \frac{PD}{4t} \quad \text{and} \quad \varepsilon_1 = \frac{\sigma}{E}$$

then $\varepsilon_1 = \dfrac{PD}{4Et}$

and with $\dfrac{\delta R}{R} = \varepsilon GF$, then

$$\frac{\delta R}{R} = \frac{PD}{4Et} GF$$

Taking the derivative of δR with respect to P yields

$$\frac{RD}{4Et} GF$$

and the static sensitivity is then

$$\frac{RD}{4Et} GF K_B \quad \text{where } K_B = \text{bridge sensitivity in V}/\Omega = 2.5$$

thus the static sensitivity for input pressure is

$$\frac{(100\ \Omega)(2\ \text{m})}{(4)(20\times 10^7\ \text{kPa})(0.02\ \text{m})} 2(2.5\ \text{mV}/\Omega) = 0.0000625\ \text{V}/\Omega \text{ or } 0.0625\ \text{mV}/\Omega$$

PROBLEM 11.18

KNOWN: A Wheatstone bridge measurement system is to be designed to measure the tangential strain in the wall of a pressure vessel. A reasonable estimate of the resulting uncertainty is desired. The bridge is to be operated in a balanced condition.

FIND: The bridge input voltage, the fixed resistance values, and the galvanometer sensitivity.

SOLUTION:

The following provides an outline for addressing this design problem. Assuming that

$R_1 = R_2 = R_3 = R_4 = 120 \ \Omega$ at balanced conditions and zero strain,

the change in R_1 with applied strain can be expressed

$$\frac{\delta R}{R} = \varepsilon GF$$

The relations for a balanced bridge are

$$\frac{R_2}{R_1} = \frac{R_4}{R_3} \text{ and with the resistances equal, } \frac{u_R}{R_1} = \frac{I_g(R_1 + R_g)}{E_i}$$

The input voltage must be designed based on a trade-off between static sensitivity from the uncertainty, and the power which must be dissipated in the bridge. The strain gauge must dissipate $I_1^2 R_1$, and should serve as the limiting factor in for the input voltage.

PROBLEM 11.19

KNOWN: It is desired to design a strain gauge based scale, using a cantilever beam. The beam is 21 cm long, 0.4 cm thick, and 2 cm wide. The loads are up to 200 g, applied 20 cm from the fixed end of the beam. The required uncertainty level is 4%.

FIND: Design measurement system, including strain gauge placement, bridge characteristics, and signal conditioning.

SOLUTION: Because this design problem has a wide variety of solutions, a general approach and some representative equations and results will be provided.

The beam is made of 2024–T4 Aluminum, having a modulus of 71 GPa. For a cantilever beam, the deflection at the free end is

$$f = \frac{Wl^3}{3EI} \qquad I = \frac{bh^3}{12}$$

where f is the deflection, W is the load, l is the distance from the fixed end to the load, I is the moment of inertia, and E is the modulus. A representative design may be examined by assuming a bridge having a single active gauge sensing the maximum axial strain. This would correspond to a location on the surface of the beam at the location where the load is applied. In this case the deflection f is

$$f = \frac{Wl^3}{3EI} = \frac{(1960)(0.2)^2}{3(71 \times 10^9)(2.67 \times 10^{-9})} = 2.7 cm \qquad I = \frac{bh^3}{12} = \frac{(0.004)(0.02)^3}{12} = 2.67 \times 10^{-9}$$

$$\sigma_x = \frac{Wl(h/2)}{I} = \frac{(1960)(0.2)(0.01)}{(2.67 \times 10^{-9})} = 1.47 \times 10^9 Pa$$

Then with E = 71 GPa

$$\varepsilon = \frac{\sigma_x}{E} = \frac{1.47 \times 10^9}{71 \times 10^9} = 0.0207 \qquad \frac{\delta R}{R} = \varepsilon GF = 0.0207 \times 2 = 0.0414$$

$$\frac{\delta E_o}{E_i} = \frac{\delta R/R}{4} = \frac{0.0414}{4} = 0.01$$

From these relationships, the output for a bridge excitation of 5 V is about 50 mV, which would create an uncertainty from the A/D resolution of about 2.4%. The bridge can be designed in a reasonable manner to meet the uncertainty constraint.

PROBLEM 12.1

KNOWN: Ranges of length to be measured

FIND: Appropriate measuring techniques

SOLUTION:

a) 0.1 to 1 m:

Tape ± 0.1 cm	Calipers (6 in. range) ± 0.0001 in.
Ruler ± 1 mm	Dial Indicator (limited range 0.5 in) ± 0.001 in.
Micrometer ± 0.001	Interferometer ± 0.0002 mm (see HP Journal, August 1970)

b) 20 to 100 A:
Scanning tunneling microscope ± 0.1 nanometer
(see *Scientific American*, October, 1989)

c) 10^6 miles:
Radio Transmissions
Laser Systems

d) 1 km
Odometer (± 0.1 miles)
Laser range finder (± 0.1 in.)

e) 100 light years
For distances to stars or galaxies comparison of brightness levels (see *Astronomy*, Vol. 17, no. 6, pg. 48)

PROBLEM 12.2

KNOWN: A steel tape with a resolution of 1/16 in (1.59 mm).

FIND: Identify sources of error in measuring the length of planking for a deck.

SOLUTION: Factors which would contribute to errors in the <u>measurement</u> include:

i) tape tension ($< \pm 0.002$ in., 0.05 mm)
ii) resolution (± 0.031 in., 0.79 mm)
iii) temperature (negligible)
iv) precision and bias errors in
scale marking   (negligible)

These errors can be combined to yield an uncertainty in a measured length as

$$u_L^{meas} = \sqrt{(0.002)^2 + (0.031)^2} = \pm 0.031 \text{ in } (0.79 \text{ mm})$$

Errors introduced in the process of cutting the planks to length would include:

i) marking errors (± 0.05 in., 1.3 mm)
ii) saw placement (± 0.1 in., 2.5 mm)
yielding a total uncertainty for the cutting process of

$$u_L^{cut} = \sqrt{(0.05)^2 + (0.1)^2} = \pm 0.11 \text{ in } (2.79 \text{ mm})$$

and for the resulting board length

$$u_L = \sqrt{(0.11)^2 + (0.031)^2} = \pm 0.12 \text{ in } 95\% \text{ (3 mm)}$$

This estimate would lead us to expect that a single plank of length 10 ft. should fall within $119.88 < L < 120.12$ in ($3.045 < L < 3.051$ m).

Comment: As in any engineering analysis, the results of the analysis should be evaluated. In the present case, engineering judgment is used in assigning magnitudes to the elemental errors. For example, an estimate of the uncertainty due to improper tension in a measuring tape can be found be examining the effect of a reasonable force on a steel tape. Letting $F_N = 10$ lb and $A_c = 0.025$ in.^{2} ($E_m = 30 \times 10^6$ psi) yields a strain of 1.33×10^{-5} and in 120 in. an error of ± 0.0016 in.(0.04 mm), which is a negligible contribution to uncertainty.

PROBLEM 12.3

KNOWN:
    a) Ruler (12 in., 30 mm)
    b) Micrometer (2 in., 5 cm)
    c) Interferometer (1 in., 2.5 cm)
    d) Calipers (4 in., 10 cm)

FIND: List possible error sources.

SOLUTION:

For each measuring instrument, error sources should be listed associated with resolution and calibration errors. Bias and precision errors in scale markings, measurement technique and temperature effects would be present for a) through c). The interferometer should have errors associated with the mirrors and the light source, as well as the means of detecting the fringe passage.

PROBLEM 12.4

KNOWN: A micrometer having a least division of 0.0001 in. (0.0025 mm) is calibrated using a gauge block set at a length of 0.1501 in. (3.8125 mm), with a limit of error for the gauge blocks of + 4 and - 2 µin (+101 and -50 µm).

FIND: The design stage uncertainty in a measurement of length.

SOLUTION:

Assuming calibration and instrument errors are the only contributing errors at the design stage, the uncertainty can be expressed as a positive and negative limit,

$$u_L^+ = +\sqrt{\left(\frac{0.0001}{2}\right)^2 + (4 \times 10^{-6})^2} \cong +0.00005$$

$$u_L^- = -\sqrt{\left(\frac{0.0001}{2}\right)^2 + (2 \times 10^{-6})^2} \cong -0.00005$$

and this yields

$$u_L = \pm 0.00005 \text{ in } (95\%) \ (0.00127 \text{ mm})$$

PROBLEM 12.5

KNOWN: A linear potentiometer having

$D = 0.1$ mm
$\rho_e = 1.7 \times 10^{-8}$ Ω-m
$R = 1$ kΩ

FIND:

a) for a core diameter of 1.5 cm, determine the range
b) plot loading error as a function of displacement

SOLUTION:

a) In order to determine the number of turns of wire, N

$$R = \frac{\rho_e L}{A_c} \Rightarrow L = \frac{RA_c}{\rho_e}$$

with

$$A_c = \frac{\pi}{4}(0.1 \times 10^{-3} \text{ m})^2 = 7.854 \times 10^{-9} \text{ m}^2$$

yields

$$L = \frac{(1000 \text{ }\Omega)(7.854 \times 10^{-9} \text{ m}^2)}{1.7 \times 10^{-8} \text{ }\Omega\text{-m}} = 462 \text{ m}$$

One turn takes $\pi(1.5)$ cm of wire, thus

$$N = \frac{462 \text{ m}}{\pi (1.5 \times 10^{-2}) \text{ m}} = 9804$$

Then since each turn occupies approximately $D = 0.1$ mm the range is
$(9804)(0.1 \text{ mm}) = 0.98$ m

b) The loading error for a voltage dividing circuit is found from (6.37) as

$$e_L = E_i \left[\left(\frac{E_o}{E_i}\right)' - \left(\frac{E_o}{E_i}\right) \right]$$

$$= E_i \frac{R_1 - R_T + (R_T - R_1)(R_1/R_m + 1)}{R_T + \left(\frac{R_T^2}{R_1} - R_T\right)(R_1/R_m + 1)}$$

In terms of output voltage the length may be written, assuming an infinite meter resistance, from (6.8)

$$E_o = \frac{L_x}{L_T} E_i = \frac{R_x}{R_T} E_i$$

yielding

$$\frac{e_L}{E_i} \times 100 = \text{loading error as a percentage of full-scale deflection}$$

A plot is shown below.

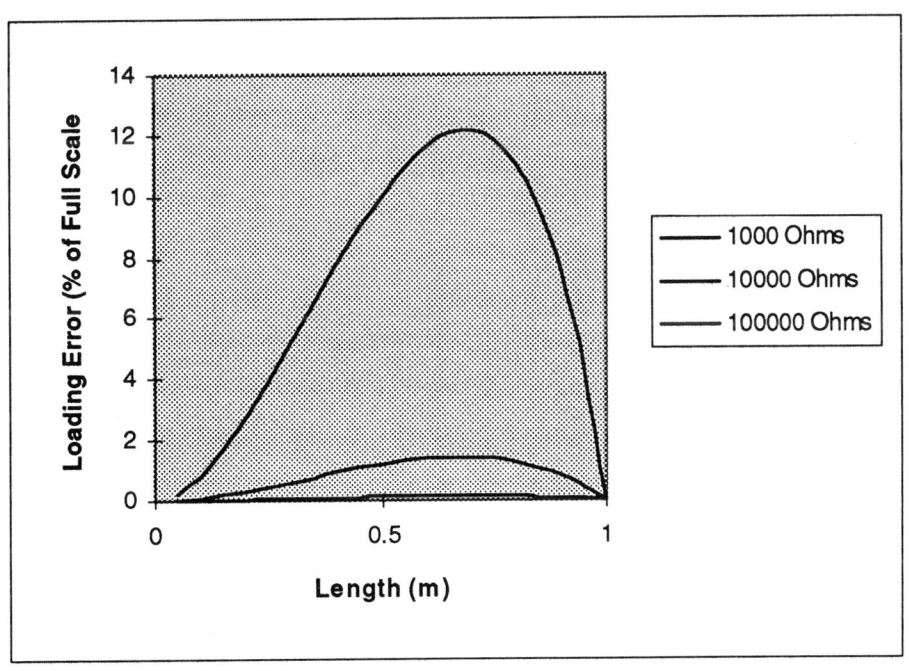

PROBLEM 12.6

KNOWN:

$l_1 = 25$ cm $\pm$ 0.5 mm
$r = 12$ cm $\pm$ 0.5 mm
$W = 350 \pm 0.1$ g
$F_u = 100$ g (nominal value)

FIND: The uncertainty in F_u.

SOLUTION:

From equation (12.5)

$$F_u = \frac{W r \sin \alpha}{l_1}$$

The desired uncertainty may be expressed

$$(u)_{F_u} = \sqrt{\sum_i \left(\frac{\partial F_u}{\partial x_i} u_{x_i}\right)^2}$$

$$= \sqrt{\left(\frac{\partial F_u}{\partial l_1} u_{l_1}\right)^2 + \left(\frac{\partial F_u}{\partial r} u_r\right)^2 + \left(\frac{\partial F_u}{\partial W} u_W\right)^2 + \left(\frac{\partial F_u}{\partial \sin \alpha} u_{\sin \alpha}\right)^2}$$

The nominal value of the angle α is

$$\sin^{-1} = \left[\frac{(100 \text{ g})(25 \text{ cm})}{(350 \text{ g})(12 \text{ cm})}\right] = 0.638 \text{ rad or } 36.5°$$

The sensitivities may be determined as,

$$\frac{\partial F_u}{\partial W} = \frac{r \sin\alpha}{l_1}$$

$$\frac{\partial F_u}{\partial r} = \frac{W \sin\alpha}{l_1}$$

$$\frac{\partial F_u}{\partial \sin\alpha} = \frac{Wr}{l_1}$$

$$\frac{\partial F_u}{\partial l_1} = \frac{-Wr \sin\alpha}{l_1^2}$$

Assuming an uncertainty of ± 0.02 rad in the angle α

$$u_{\sin\alpha} = \frac{\partial \sin\alpha}{\partial \alpha} u_\alpha = (\cos\alpha) u_\alpha = \pm 0.016 \text{ rad}$$

$$u_{F_u} = \left[\left(\frac{-350(12)(0.595)}{(25)^2} 0.05 \right)^2 + \left(\frac{-350(0.595)}{(25)} 0.05 \right)^2 + \left(\frac{(12)(0.595)}{(25)} 0.1 \right)^2 + \left(\frac{350(12)}{(25)} 0.016 \right)^2 \right]^{\frac{1}{2}}$$

$$= [0.04 + 0.17 + 8.2 \times 10^{-4} + 7.23]^{\frac{1}{2}}$$
$$= \pm 2.73 \text{ g } (95\%)$$

PROBLEM 12.7

KNOWN:

$y(t) = 0.2 \cos 10t + 0.3 \cos 20t$
where y = displacement [in.]
and t = time [sec.]
$\zeta = 0.7$
$k = 1.2$ lb/ft

FIND:

a) a combination of m and c which would yield less than 10% amplitude error in measuring the input signal.
b) determine the phase response of the system

SOLUTION: We know

$$\omega_n = \sqrt{k/m} \qquad c_c = 2\sqrt{km}$$

The amplitude error is evaluated by examining (for cos = 1)

$$\frac{(y_r)_{steady}}{A} = \frac{(\omega/\omega_n)^2}{\left\{\left[1-(\omega/\omega_n)^2\right]^2 + \left[2\zeta(\omega/\omega_n)\right]^2\right\}^{1/2}}$$

The limiting case is for the lower input frequency, and $(y_r)_{steady}/A = 0.9$

$$0.9 = \frac{(10/\omega_n)^2}{\left\{\left[1-(10/\omega_n)^2\right]^2 + \left[2(0.7)(10/\omega_n)\right]^2\right\}^{1/2}}$$

Solving for ω_n yields

$$\omega_n = 7.1 \text{ rad/s}$$

and since

$$\omega_n = \sqrt{\frac{kg_c}{m}}$$

and c is found from

$$c = 2\zeta\sqrt{km/g_c}$$

yielding

$$c = 2(0.7)\sqrt{\frac{1.2 \text{ lb/ft}(0.766 \text{ lbm})}{32.174 \frac{\text{ft-lbm}}{\text{lb-sec}^2}}} = 0.236 \frac{\text{lb-sec}}{\text{ft}}$$

b) Phase response is shown below.

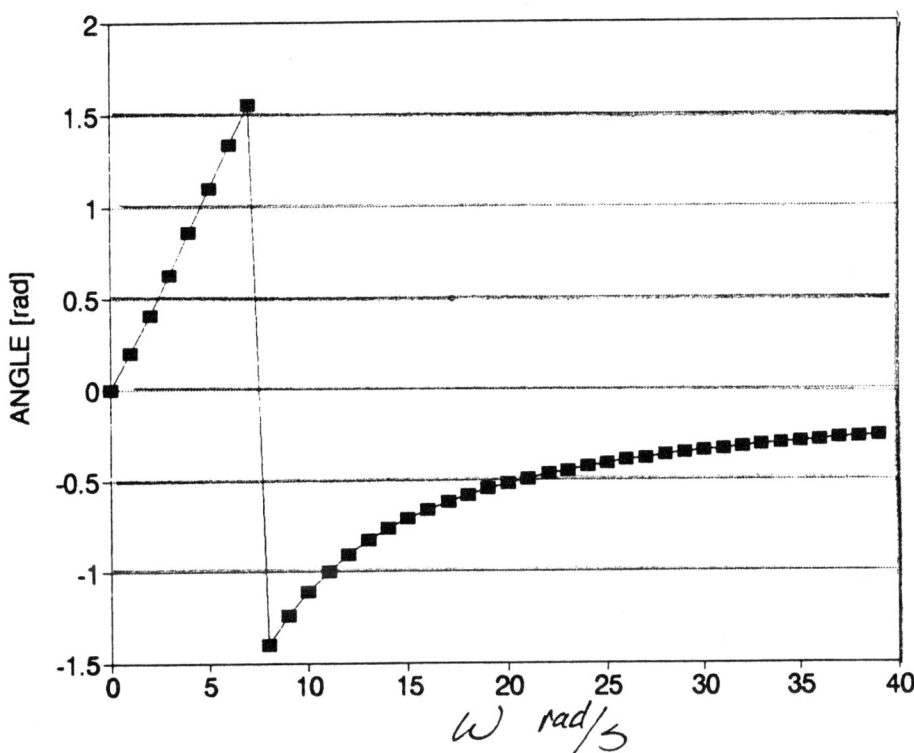

PROBLEM 12.8

KNOWN: A seismic instrument has
$\omega_n = 20$ Hz $= 40\pi$ rad/s $\zeta = 0.65$

FIND: The maximum input frequency, ω, for a vibration measurement such that the amplitude error is $< 5\%$.

SOLUTION:
From (12.14)

$$0.95 \text{ or } 1.05 = \frac{1}{\left\{\left[1-\left(\omega/\omega_n\right)^2\right]^2 + \left[2\zeta\left(\omega/\omega_n\right)\right]^2\right\}^{1/2}}$$

Solving for the input frequency yields 90.5 rad/s.

PROBLEM 12.9

KNOWN: A seismic accelerometer has:
m = 0.2 g
k = 20 000 N/m
Very low damping

FIND: Instrument bandwidth

SOLUTION:

The natural frequency is found as

$$\omega_n = \sqrt{k/m} = \sqrt{\frac{20{,}000 \; \frac{\text{kg-m}}{\text{sec}^2\text{-m}}}{0.2 \times 10^{-3} \; \text{kg}}} = 10{,}000 \text{ rad/sec}$$

and with a very low damping. The bandwidth may be found from Figure 3.16 with the allowed frequency range from 0 to 0.4 (ω_n) or 0 to 4000 rad/s.

PROBLEM 12.10

KNOWN: Integration reduces the effects of noise in a signal
A moving average integrates over a fixed time interval, based on a concept called windowing
Noise in the present case has significant amplitude, but a significantly higher frequency than the measured velocity

SOLUTION:

Consider a signal having the stated characteristics. An example is shown below.

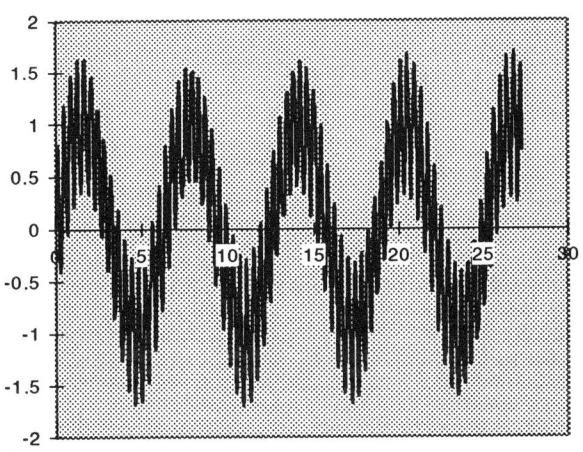

This signal has high frequency noise present

A moving average has been performed on the signal below! It is necessary to average over several periods of the noise to achieve the desired result.

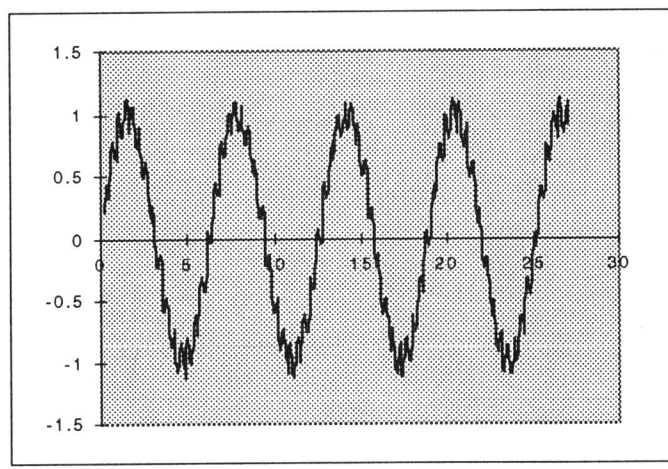

PROBLEM 12.11

KNOWN: Moving coil transducer with
$D_c = 0.8$ cm $l = 2$ cm dy/dt from 1 to 10 cm/s
A/D converter with 8 bit resolution and 0.1% FS accuracy

FIND: The required magnetic field strength as a function of N for 1% accuracy in the velocity measurement.

ASSUMPTION: Assume that the uncertainty in the magnetic field strength and the number of turns are negligible.

SOLUTION:

From equation 12.19, the emf from the moving coil can be expressed,

$$emf = \pi B D_c l N \frac{dy}{dt}$$

the uncertainty in the velocity can then be expressed, with $V = dy/dt$

$$\frac{u_V}{V} = \frac{u_{emf}}{emf}$$

The uncertainty in the emf has two contributions. From the resolution of the A/D, the uncertainty is ±7.8 mV plus 1 mV (RSS addition) from the accuracy, yields u_{emf} = 7.9 mV. Combining the relations for emf and the uncertainty in V, with an uncertainty in V of 1%, yields the plot below.

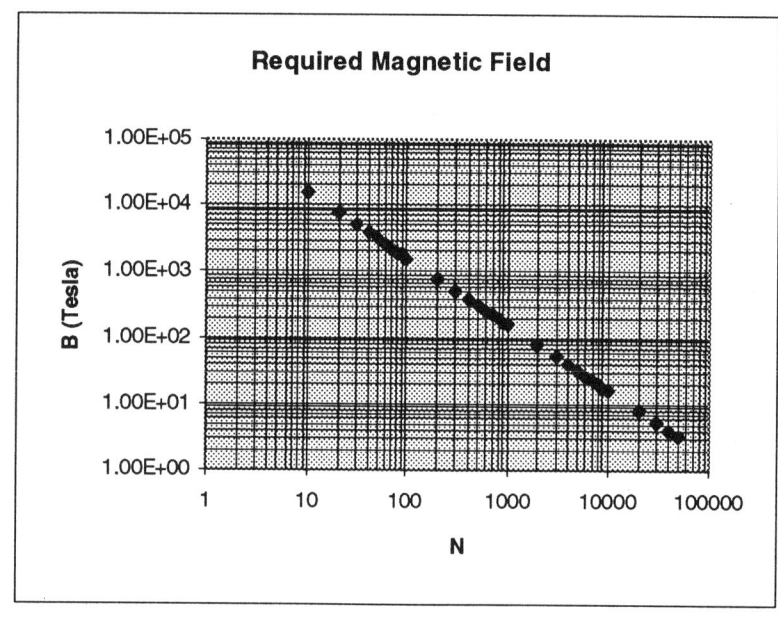

PROBLEM 12.12

KNOWN: Rotational speed measured by a stroboscope. The measured rotational speed is higher than the flash rate. The strobe synchronizes at 10,000, 18,000 and 22,000 RPM

FIND: Rotational speed.

SOLUTION: From Equation (12.20)

$$\omega = \frac{\omega_1 \omega_N (N-1)}{\omega_1 - \omega_N} = \frac{(22{,}000)(10{,}000)(3-1)}{(22{,}000 - 10{,}000)} = 36{,}666 \text{ RPM}$$

PROBLEM 12.13

KNOWN: A proving ring is to be designed to serve as a calibration standard for forces over the range from 250 to 1000 N.

FIND: A suitable design to provide reasonable uncertainty.

SOLUTION: It is necessary to establish the minimum deflection which can be measured with reasonable accuracy. As an example, consider a displacement transducer having a range of 0 to 1 mm for an output of 0 to 5 Volts. If sampled using an 8-bit A/D, the resolution would be 19.5 mV, corresponding to 0.004 mm. If we assumed that the proving ring would deflect 1 mm at 1000 N, we can size the ring.

Assuming the cross section of the ring is rectangular, the moment of inertia is $bh^3/12$, and the deflection is given by

$$\delta y = \left(\frac{\pi}{2} - \frac{4}{\pi}\right) \frac{F_n D^3}{16 EI}$$

Let's assume that the ring is steel with a modulus of $E = 20 \times 10^{10}$ Pa. By varying the cross section and the diameter, a suitable deflection can be established. As an example of the process, assume a square cross-section, and a diameter of 8 cm. A deflection of 1mm at 1000 N would be achieved with a dimension of 4.9 mm for the square cross-section. Then at a load of 250 N, the deflection would be 0.25 mm, yielding an output voltage of 1.2 V, which should yield a reasonable uncertainty.

PROBLEM 12.14

KNOWN: Power transmission through a drive shaft results in 1800 rpm with a power transmission of 40 HP.

FIND: Torque transmitted by the driveshaft.

SOLUTION:
With

$$P = \omega T$$

and $P = 40$ HP with $\omega = \dfrac{1800 \times 2\pi}{60} = 188.5$ rad/s

$$T = \frac{(40 \text{ HP})(550 \text{ ft-lb/s-HP})}{188.5 \text{ rad/s}} = 117 \text{ ft-lb}$$